THE INTERNATIONALIZATION OF TECHNOLOGICAL INNOVATION FOR CHINESE ENTERPRISES

Series on Innovation and Operations Management for Chinese Enterprises

Print ISSN: 2591-7188
Online ISSN: 2591-7196

Editor: Xiaobo Wu *(School of Management, Zhejiang University, China)*

Published:

Vol. 3: *The Internationalization of Technological Innovation for Chinese Enterprises*
by Jin Chen

Vol. 2: *Basic Research and Industrial Innovation in China*
by Xielin Liu, Yubing He and Peng Cheng

Vol. 1: *Independent Innovation in China: Theory and Cases*
by Huibo Zhong and Jiasu Lei

Forthcoming:

Vol. 4: *The Theory and Policies of Mutual Benefit and Win-Win Strategy: Research on Sustainable Development of China's Open Economy*
by Lixin Yu and Wanling Chen

Series on Innovation and Operations Management for Chinese Enterprises – Vol. 3

THE INTERNATIONALIZATION OF TECHNOLOGICAL INNOVATION FOR CHINESE ENTERPRISES

Jin Chen
Tsinghua University, China

World Scientific

ZHEJIANG UNIVERSITY PRESS
浙江大学出版社

Published by

World Scientific Publishing Co. Pte. Ltd.
5 Toh Tuck Link, Singapore 596224
USA office: 27 Warren Street, Suite 401-402, Hackensack, NJ 07601
UK office: 57 Shelton Street, Covent Garden, London WC2H 9HE

and

Zhejiang University Press
No. 148, Tianmushan Road
Xixi Campus of Zhejiang University
Hangzhou 310028, China

Library of Congress Cataloging-in-Publication Data
Names: Jin, Chen, 1968– author.
Title: The internationalization of technological innovation for Chinese enterprises / Jin Chen.
Description: New Jersey : World Scientific, [2018] | Series: Series on innovation and operations management for Chinese enterprises ; vol 3
Identifiers: LCCN 2017056374 | ISBN 9789813208667
Subjects: LCSH: Technological innovations--Economic aspects--China. | Industries--Technological innovations--China.
Classification: LCC HC430.T4 J556 2018 | DDC 338/.0640951--dc23
LC record available at https://lccn.loc.gov/2017056374

British Library Cataloguing-in-Publication Data
A catalogue record for this book is available from the British Library.

This edition is jointly published by World Scientific Publishing Co. Pte. Ltd. and Zhejiang University Press. This edition is distributed worldwide by World Scientific Publishing Co. Pte. Ltd., except China.

For any available supplementary material, please visit
https://www.worldscientific.com/worldscibooks/10.1142/10411#t=suppl

Typeset by Zhejiang University Press

Printed in China

Preface

Technological innovation is an activity with collaborative efforts to fully use the external networks and resources of enterprises to create, expand, and enhance their capabilities. It enables the formation of an enormous knowledge transaction network in an enterprise and between the enterprise and various external organizations. With such a network, the enterprise is able to use a broader range of external knowledge and information to build its own innovation framework. Compared with the traditional technological innovation mode, the internationalization of technological innovation represents a paradigm shift in technological innovation.

Internationalization of technological innovation has already become the dominant design and winning weapon for business enterprises to effectively carry out technological innovation, enhance technological capabilities, and engage in international competition. Faced with enormous competitive pressure from large international companies with strong technological power, Chinese enterprises must raise the foothold of their technological innovation strategies and speed up the implementation of internationalized strategy for technological innovation, in order to keep up with the pace of worldwide technological innovation and attain strategic successes in both national-level and enterprise-level technological innovation.

Based on a new perspective, this book explores the theory of internationalization of technological innovation in four aspects, namely, strategy, organizational form, resource allocation, and environment, and analyzes the development rules and key points of management in internationalization of technological innovation for Chinese enterprises. On the basis of theoretical and empirical studies, this book proposes an evolutionary model of internationalization of technological innovation. It is hoped that the proposed model will be of value to the advancement of internationalization of technological innovation for Chinese enterprises and will contribute to the enrichment and improvement of theories in technological innovation.

There are seven chapters in the main body of this book. Chapter 1 analyzes the development trends and characteristics of internationalization of technological innovation, demonstrates the necessity for Chinese enterprises to implement an internationalized strategy in technological innovation, and introduces the research design of this book. Chapter 2 reviews the existing theories on internationalization of technological innovation and presents brief comments. Chapter 3 studies the theories on internationalization of technological innovation and proposes a strategic framework of internationalization of technological innovation for Chinese enterprises that is based on core competency as well as strategic approaches to internationalization of technological innovation. Chapter 4 studies organizational management theories on internationalization of technological innovation from three aspects, namely, organizational structure, organizational exchange and communication, and technological innovation process, and proposes a three-phase mode of organizational evolution in the process of internationalization of technological innovation for Chinese enterprises. Chapter 5 studies theories of resource allocation for internationalization of technological innovation of enterprises and proposes a resource allocation model. Chapter 6 analyzes the influence mechanism of environmental factors on internationalization of technological innovation for Chinese enterprises and puts forward some policy suggestions. Chapter 7 summarizes the previous chapters with regard to aspects such as strategy, organization, resource allocation, and environment and proposes the evolutionary model of internationalization of technological innovation for Chinese enterprises.

In addition, as provided in the appendices, we have conducted several typical case studies of enterprises on internationalization of technological innovation and summed up the characteristics of internationalization of technological innovation in the enterprises, which were in different development stages of the internationalization process of technological innovation. There is also a questionnaire for China's pioneer enterprises in research and development (R&D) internationalization.

This book is a reorganization and refinement of the research report of the project, "Research on the Internationalization of Technological Innovation for Chinese Enterprises in Response to China's Accession into the WTO," which was funded by the Director's Fund of the National Natural Science Foundation of China.

In the process of conducting the questionnaire survey and interview, we have obtained support from many leading domestic companies and would like to express our deep gratitude to all of them.

During the project's research process and book writing process, famous scholars from China and abroad, such as Professor Maximilian Von Zedtwitz of the International Institute for Management Development (IMD), Switzerland, Professor Xu Qingrui, Director of Zhejiang University Innovation and Development Research Center, Professor Fang Xin of the Chinese Academy of Sciences, and Professor Xue Lan from the School of Public Administration of Tsinghua University, have given their attention. Thanks go to all of them.

During the project's research process, Professor Zhu Zhaohui, Dr. Jing Jinsong, Dr. Wu Canglan, Dr. Tong Liang, Dr. Zhang Hongshi, and others have done some important work—my sincere gratitude to all of them.

Because of my limited knowledge and abilities, deficiencies in this book cannot be avoided. I would welcome comments from readers, and I am sure the comments/critiques will enrich my understanding.

Jin Chen
Professor, School of Economics and Management,
Tsinghua University
December 2017

Contents

Chapter 1 Introduction 1

1.1 Background and significance of research 3

1.2 Ideas and methods 12

1.2.1 Research ideas 12

1.2.2 Research methods 18

References 21

Chapter 2 Internationalization of Technological Innovation: A Literature Review 23

2.1 Determinants of internationalization of technological innovation 25

2.2 Types of internationalization of technological innovation 30

2.3 Location choices of internationalization of technological innovation 36

2.4 Development process and organizational form of internationalization of technological innovation 41

2.5 Management of internationalization of technological innovation 46

2.5.1 Centralization and decentralization 47

2.5.2 Integration of internationalized R&D activities 49

2.5.3 Exchange and communication 51

2.5.4 Human resource management 53

2.6 Chapter summary 55

References 58

Chapter 3 Internationalization of Technological Innovation: Strategic Framework and Strategic Approach 65

3.1 Theories of technological innovation and technological innovation strategy—origin of internationalization strategy of enterprise technological innovation 67

3.2 Analysis of the state of Chinese enterprises on internationalization of technological innovation 77
3.2.1 Scale and distribution of R&D funding and types of internationalized innovation activities 78
3.2.2 Strategic factors for implementing overseas technological innovation 81
3.3 Strategic framework of Chinese enterprises for internationalization of technological innovation 84
3.4 Strategic approaches of internationalization of technological innovation for Chinese enterprises 92
3.4.1 Strategic Approach A: domestic R&D for domestic market with technology monitoring-oriented internationalization 93
3.4.2 Strategic Approach B: domestic R&D for domestic market first and then for international market with the purpose of international cooperation and technology monitoring 95
3.4.3 Strategic Approach C: domestic R&D branches mainly for domestic market, with overseas R&D branches established directly by the enterprise or obtained through M&As 97
3.4.4 Strategic Approach D: international R&D for international market 100
3.5 Chapter summary 102
References 104

Chapter 4 Internationalization of Technological Innovation: Organization and Management **107**
4.1 Theoretical study on organization and management of internationalization of technological innovation 109
4.1.1 Organizational forms of internationalization of technological innovation 110
4.1.2 Centralization and decentralization 113
4.1.3 Exchange and communication 115
4.1.4 Integration of internationalization of technological innovation 118
4.2 Analysis of the organization and management of internationalization of technological innovation for Chinese enterprises 121
4.2.1 Organizational form 121
4.2.2 Exchange and communication 127
4.2.3 Integration of internationalized innovation processes 134
4.3 Suggestions on internationalization of technological innovation for Chinese enterprises 144
4.3.1 Evolution mechanism of the organizational form 144

4.3.2 Exchange and communication 154
4.3.3 Integration of internationalized innovation processes 162
4.4 Chapter summary 169
References 173

Chapter 5 Internationalization of Technological Innovation: Resource Allocation and Integration 175

5.1 Capital investment and allocation for internationalization of technological innovation 178
5.1.1 Capital investment and allocation for internationalization of technological innovation 178
5.1.2 Status quo of capital investment and allocation of China's internationalization of technological innovation 187
5.2 Allocation of human resources for internationalization of technological innovation 191
5.2.1 Selection and acquisition of human resources for internationalization of technological innovation 191
5.2.2 Human resources for China's internationalization of technological innovation 197
5.3 Acquisition, development, and transfer of technological information sources for internationalization of technological innovation 203
5.3.1 Acquisition, development, and transfer of technological information sources 203
5.3.2 Analyzing the present state of technological information sources of China for internationalization of technological innovation 208
5.4 Effective resource allocation and integration in the process of internationalization of technological innovation for Chinese enterprises 211
5.4.1 Comprehensive analysis of models for directing innovation resource flow in the initial stage of internationalization 211
5.4.2 Effective allocation and integration of innovation resources in China's internationalization of technological innovation 219
5.5 Chapter summary 228
References 230

Chapter 6 Internationalization of Technological Innovation: Environmental Factors 233

6.1 Elements and characteristics of environmental factors for internationalization of technological innovation 236

6.1.1 Environmental factors for enterprises' internationalization of technological innovation 236
6.1.2 Environmental characteristics of enterprises' internationalization of technological innovation 238
6.2 Impact of host country's environmental factors on enterprises' internationalization of technological innovation 240
6.2.1 Impact of host country's environmental factors on internationalization of technological innovation that can result in location advantages or location disadvantages 240
6.2.2 Mechanism by which environmental factors of host country affect enterprises' internationalization of technological innovation 244
6.3 Impact of home country's environmental factors on enterprises' internationalization of technological innovation 254
6.3.1 Mechanism by which policies of home country affect enterprises' internationalization of technological innovation 254
6.3.2 Mechanism by which innovative resource environment of home country affects enterprises' internationalization of technological innovation 257
6.3.3 Mechanism by which industry organization environment of home country affects enterprises' internationalization of technological innovation 259
6.4 Analysis of environmental factors and policies of internationalization of technological innovation for Chinese enterprises 263
6.4.1 Analysis of the effect of host country's environmental factors on internationalization of technological innovation for Chinese enterprises 264
6.4.2 Analysis of the impact of government policy adjustment on enterprises' internationalization of technological innovation upon China's accession to the WTO 274
6.4.3 Policy suggestions 276
6.5 Chapter summary 280
References 282

Chapter 7 Conclusion **285**
7.1 Strategic aspect of internationalization of technological innovation 287
7.2 Organizational aspect of internationalization of technological innovation 291
7.3 Resource allocation aspect of internationalization of technological innovation 293

7.4 Managerial aspects of internationalization of technological innovation 295
7.5 Evolutionary modes and management features of China's internationalization of technological innovation 297
7.6 Environmental support conditions of internationalization of technological innovation for Chinese enterprises and policy implications 300

Appendix 1 303

Case 1 Motorola's innovation internationalization 305
1. Profile of Motorola China 306
2. Technology transfer and localization—main R&D center 307
3. Technology transfer and localization—localization of ancillary products 312

Case 2 Haier's internationalization of technological innovation: inexhaustible source of power 316
1. Technological innovation: Haier going from the brink of bankruptcy to the realm of glory 316
2. Internationalization of technological innovation: Haier, going from China to the world 321
3. Revelation for Chinese enterprises from Haier's internationalization of technological innovation 325

Case 3 Holley Group's internationalization of technological innovation 328
1. Chance 329
2. Competition and cooperation with Qualcomm 331
3. Reflection 333

Case 4 Huawei's internationalization of technological innovation 335
1. Company profile 337
2. Huawei's strategy for internationalization of technological innovation 338
3. Organizational structure and resource allocation of Huawei's internationalized technological innovation system 346
4. Revelation of Huawei's internationalization of technological innovation 351

Case 5 GE's internationalization of technological innovation: experience and what it teaches us 354
1. Historical path of GE's internationalization of technological innovation 355
2. Internationalization process of GE's technological innovation 359
3. Experience and meanings 367

Case 6 Wanxiang Group's internationalization 370
1. Wanxiang's history of development 370
2. Wanxiang's management structure 372
3. Wanxiang's internationalization process 374
4. Wanxiang's internationalization of technological innovation 377

Case 7 ZTE's internationalization of technological innovation 386
1. Internationalization of innovative sources 388
2. Establishing technology alliances and achieving complementary advantages 391
3. Going abroad and setting up overseas R&D institutions 393
4. Hybrid management method combining top-down method and bottom-up method 396
5. Road of internationalization of technological innovation for Chinese enterprises as seen in ZTE's experience 400

Case 8 3M's global innovation 403
1. About 3M 403
2. Global innovation—3M's advantages 405
3. 3M's global innovation management 413

References 429

Appendix 2 433

Questionnaire for China's pioneer enterprises in R&D internationalization 435

Index 455

Chapter 1

Introduction

1.1 Background and significance of research

Ever since the U.S. economist Joseph Schumpeter proposed the idea that "innovation is to introduce into the production system 'a combination of production factors' that has never existed," innovation has played a crucial role in a country's economic development, especially in business competition, in order to grab the attention of a greater number of people.

Currently, the breakneck speed of advancement in science and technology (S&T) and the globalization of world economy are two prominent features in the development of human society. The explosive growth in scientific and technological knowledge has made knowledge the most dominant element in bringing about economic growth and profit creation. Competitive enterprises—through developing, protecting, improving, and updating knowledge and by continued technological innovation—rapidly and continually push onto the market products that have been created or improved with knowledge and, as a result, gain lasting competitive advantages; yet, the globalization of world economy means that the markets worldwide are being integrated into one and, as a result of such change, competition among enterprises has turned domestic competition into international competition. Internationalization of business operations has further intensified international competition. Under such circumstances, enterprises that have a global vision and are capable of consolidating worldwide resources and innovating rapidly will win this competition.

Development in technology has made technological innovation—which turns technological knowledge into practical productivity—the key factor in gaining competitive advantages; meanwhile, globalized competition has made globally acquiring and allocating innovative resources the key factors in building up business capabilities. Adapting to this situation, enterprises in the industrialized nations worldwide all regard the realignment of productive factors arising from technological innovation as the main source of enterprise development. Especially, since the 1980s, with the rapid development of economic globalization and the ever-intensifying international competition, large multinationals in advanced countries—to meet the demands arising from the complexity of the markets worldwide, the wide variety of products, and the different preferences of consumers of different countries, to make use of the special technological resources of various countries worldwide; and to contain the costs and risks involved in the process of new product development—walked away from the singular mode of home-country-based technological innovation. In accordance with the comparative advantages in talent and technological strength and in technological infrastructure of host countries, they have started to systematically expand their technological innovation activities across the globe, thereby pushing technological innovation in transnational corporations to be ever more internationalized and globalized (Lin & Chai, 1998).

At the macro level, according to "The Globalization of Industrial Research and Development" report published by the U.S. Department of Commerce, between 1987 and 1993, the total investment abroad in R&D by U.S. companies increased from \$5.2 billion to \$9.8 billion. Such investments are not forever concentrated on developed countries only; the growth rate of such investments in R&D done in developing countries is actually very high. From 1989 to 1993, the total investment in R&D done in Ireland by U.S. companies increased from \$134 million to \$669 million; in Brazil, from \$940,000 to \$220 million; and in Spain, from \$115 million to \$320 million. However, in Singapore, the total investment by U.S. companies jumped from \$5 million to \$312 million, an astonishing pace of growth (Wang, 1997). At the micro level, many companies have set up institutions overseas to conduct R&D activities abroad. In 1987, 9.3% of all business-funded R&D activities were conducted by foreign companies; in 1995, this figure rose to nearly 18%. There are 60 Canadian companies that are operating nearly 100 research institutes in the United States, Western Europe, Japan, Australia, and in some developing countries. The portion of R&D investment that was done abroad by Japanese companies grew from 1.58% of the total in 1987 to 5% in 1991. The portion of R&D investment that was done abroad by Swiss companies grew from 15% of the total in 1987 to 23% in 1991, which put the annual growth rate at 32% (Niosi, 1999).

Internationalization of technological innovation is for an enterprise to extend its innovative activities abroad—through mergers and acquisitions (M&As), by directly establishing overseas R&D institutions, by forging international technology alliances, or by other means; it is also a new paradigm of technological innovation characterized by globalized resource acquisition, internationalized talent for innovation, and networked innovative organizations. From the point of view of information acquisition, the innovative sources of the internationalization of technological innovation are globally available—an enterprise can acquire innovative thoughts and information from a wider range of channels. From the point of view of innovation strategy, the purpose of internationalization of technological innovation is to take advantage of the diverse technological resources of various countries and to harness both the internal and external forces to the organization in order to enhance its technological capabilities. From the point of view of organizational form, internationalization of technological innovation transforms the organizational structure of the innovative function of an enterprise into a global collaborative network. From the point of view of resource input scheme, the internationalization of technological innovation makes it possible to make better use of the high-quality and low-cost human resources of various countries, better environment for technological innovation, and more advanced infrastructure for R&D.

Traditional technological innovation is set inside an enterprise; it is to conduct innovative activities based on the existing resources of the enterprise. This vision of innovation is indeed narrow and small; it did not include innovation strategies in either the connectedness of the enterprise and its external environment or the acquisition and allocation of resources from a wider range of sources. Different from traditional forms of technological innovation, internationalization of technological innovation starts with the enterprise having a full view of the globe, which enables the enterprise to work through collaborative mechanisms with outside entities to acquire and allocate worldwide resources in an effort to break through the resources and capability constraints found inside the enterprise itself and to take full advantage of the external networks and resources to build, expand, and enhance the business capabilities. It also enables the formation of an enormous knowledge transaction network connecting the internal system of the enterprise with various external organizations and changes the situation when the enterprise can only obtain necessary technological knowledge internally. Therefore, the enterprise will be able to draw from a broader range of external knowledge and data sources to build up its own technology and intelligence system. In view of the traditional mode, internationalization of technological innovation represents a paradigm shift—it is a whole new model of technological innovation.

After years of field practice and accumulated experience, transnationals of the world's major industrialized countries have learned many winning courses of action in internationalization of technological innovation. They have formulated systematic approaches and advanced models in areas such as the choice of the strategic path to attain internationalization of technological innovation, the choice of the form of internationalization, the selection of the regions and sites to establish R&D centers in the host countries, the establishment of organizational structures, the integration of resources, and so on.

These accumulated experience and effective field work for internationalization of technological innovation have provided transnational corporations with powerful support in their international expansion. First, the market-support-type internationalization of technological innovation—which is done in response to the various special needs of the varied customers worldwide and in adaptation to the conditions of local production and is done to achieve the localization of technology—has strongly encouraged the production operations of transnationals in their host countries and has become the means to better raise competitiveness on the basis of local markets and local conditions of production. Next, the technology-search-and-acquisition-type internationalization of technological innovation—which is done to open up the windows of information, to track and acquire technology from the host countries and competitors, to get close to the knowledge centers, and to benefit from the spillover effect of the local R&D activities—has provided transnationals with additional sources of various technologies and has become an important way for them to effectively promote their technological capabilities. Furthermore, the resource-utilization-type internationalization of technological innovation—which is done to take advantage of the technological talent and research environment of the host country and to reduce the R&D costs—has become an important approach for transnationals to have the risks and costs involved in innovation and to raise its efficiency. It can be said that internationalization of technological innovation is already the dominant design and winning weapon for business enterprises in the developed countries to effectively carry out technological innovation, to enhance technological capabilities, and to engage in international competition.

Now, China has acceded to the WTO; this means that the domestic market and the international market are becoming integrated, that the domestic supply and demand will no longer be the primary deciding factors of the rise and fall of an enterprise, and that the national economic development and the fate of each enterprise will be rearranged on the larger stage of the world economy; this also means that our business enterprises will have to follow the unified rules of market economy when participating in the international economic competition. Generally, Chinese business enterprises are weak in many aspects, operating at a low level of internationalization and under intense competitive pressure from internationally operated big businesses that enjoy established and strong technological capabilities. How to meet the demands of a global economy and the internationalized competition and how to build up and enhance core capabilities through effective technological innovation so as to be able to participate in international economic and technological cooperation and competition at a higher level in a broader area and a wider field are now important questions faced by Chinese enterprises.

In China, the status of business enterprises as the locus for technological innovations has been gradually established in the process of building a market-oriented economic system. Compared with those in the developed countries, business enterprises in China have not been operating for the market and performing as the locus of technological innovation for long and, thus, inside these enterprises, the resources allocated for technological innovation are generally lacking. In the national innovation system of China, the components, regardless of the government units, enterprises, universities, or research institutes, no matter viewed horizontally or vertically, are too loosely related to each other with respect to technological innovation. Far from being able to form an organic and collaborative whole system (Yang & Bai, 2000), technological innovations by the individual enterprises are still not effectively supported by China's national innovation system and technological innovations that are set internally and done behind closed doors are relatively common. However, there is a rather big gap between the scientific and technological strength of China and that of the developed countries worldwide, and the latest scientific and technological knowledge developed in the companies is mostly of foreign origin. For these reasons, the innovative capabilities of Chinese enterprises in various fields of technology are underdeveloped.

With the field of view limited to the enterprise itself and the reliance placed only on existing resources, traditional technological innovation will only make the technological capabilities of Chinese enterprises lag more and more behind those in the developed countries. Only by upgrading the strategic position of technological innovation, speeding up the implementation of the strategy for internationalization of technological innovation, constructing the necessary external networks, taking advantage of both domestic and foreign resources, and broadening the ways to acquire and create capabilities, can the Chinese enterprises catch up with the pace of worldwide technological development and attain strategic successes in both national-level and enterprise-level innovation; only then can these enterprises build up and strengthen their core capabilities and gain a stronger foothold to participate in international competition.

From a realistic perspective, at this stage, most Chinese enterprises are limited by their own individual strengths and the national conditions and thus are basically in a blank or early stage with respect to internationalization of technological innovation, have not acquired an in-depth understanding of the various aspects of internationalization of technological innovation, and lack the necessary experience for executing internationalization of technological innovation. Thus, the need for a theory is compelling—which should meet the actual conditions of Chinese enterprises and provide guidance for internationalization of their technological innovation.

In theoretical studies, foreign scholars—from the perspective of multinationals—have conducted a lot of research on motivation, strategy, organization, management, etc. for internationalization of technological innovation. However, such research on internationalization of technological innovation by foreign scholars was mainly aimed at the enterprises of developed countries; research targeted at the enterprises in developing countries such as China is still basically in a blank state. Compared with transnationals, enterprises in the developing countries operate entirely at different levels of strength and are faced with a very different competitive environment; hence, their paths toward internationalization of technological innovation will necessarily be very different from those of the transnationals. Research on internationalization of technological innovation for enterprises of developing countries is an important aspect of the theoretical studies on internationalization of technological innovation. Without a theory on internationalization of technological innovation for enterprises of developing countries, the system of theories in this field remains incomplete.

So far, there have been no systematic studies on internationalization of technological innovation for the enterprises of China. Carefully analyzing internationalization of technological innovation for the technologically-advanced transnational corporations, drawing on the studies done abroad on such internationalization, systematically studying the theory for internationalization of technological innovation for the Chinese business enterprises, shaping the operational model and management approach for the internationalized technological innovation suitable for Chinese enterprises, and pushing for theoretical advancement beyond existing research have not only great theoretical meaning but also great practical significance.

It is in this context that we have been engaged in the study on internationalization of technological innovation for Chinese enterprises. This project is about the theories of technological innovation; we will start from a systematic study and comprehensive analysis of the state of internationalization of technological innovation in the transnationals of industrialized countries; stay focused on the actual situations faced by China after accession to the WTO; strive to shape the theory, operational model, and management approach of internationalization of technological innovation that are suitable for China; and seek to find effective ways for Chinese business enterprises to internationalize technological innovation and a new form to manage such technological innovation.

1.2 Ideas and methods

1.2.1 Research ideas

Transnational corporations are the ones that started and have been advancing internationalization of technological innovation. During the period from 1985 to 1995, setting up transnational R&D institutions was a powerful trend (Gerybadze & Reger, 1999). Basically, the research on internationalization of technological innovation was gradually developed alongside the development of internationalization of technological innovation by the transnational corporations.

For the present state of the research on internationalization of technological innovation abroad, the objects of these studies are mainly transnationals, the focus is mainly on internationalization of technological innovation for enterprises in developed countries, and studies aimed at business enterprises in developing countries are basically nonexistent. Studies are mostly done from the perspective of the traditional form of the management of transnational corporations and, therefore, have neglected the flexibility and dynamism of the new organizational forms and new systems. Studies targeted at the new trend of globalized innovation, that is, internationalization of technological innovation with an emphasis on the creation of knowledge and capabilities—are lacking. With respect to strategy research, the focus has been on the internal motivation for internationalization of technological innovation and the determinant factors of region and site selection, and the theoretical and framework studies for strategy analysis with the goal of boosting the technological capabilities of business enterprises are rarely involved. In addition, systemic studies on internationalization of technological innovation from the perspective of balance and agreement among its organizational structure and processes are relatively less.

With our study, internationalization of technological innovation is for a business enterprise to have a full view of the globe, to work through collaborative mechanisms with outside entities to acquire and allocate resources worldwide in an effort to break through the resources and capability constraints found inside the enterprise itself, and to take full advantage of the external networks and resources for the enterprise to build, expand, and enhance business capabilities. In view of the traditional mode of technological innovation, it represents a shift in paradigm (see Figure 1-1 for the SPRE model: Strategy—Process—Resource—Environment).

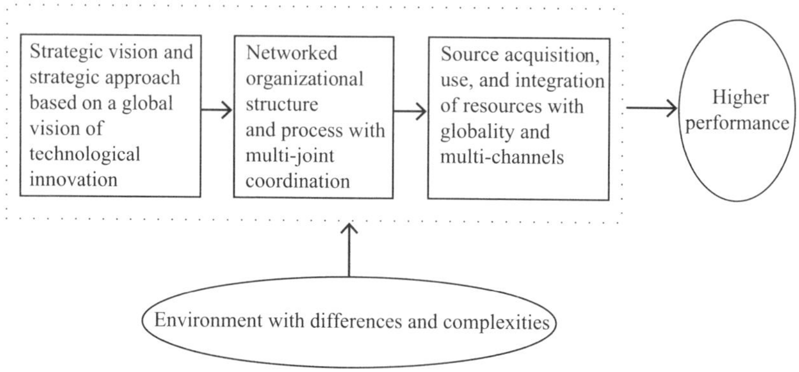

Figure 1-1 The SPRE model of internationalization of technological innovation

Internationalization of technological innovation extends innovative activities overseas. Compared with the past technologically innovative activities that were restricted to the home country alone, geographically dispersed R&D activities are influenced not only by the environment of the home country but also by that of the host countries. Hence, the environment for technological innovation faced by business enterprises will be characterized by greater diversity, complexity, and variety. The differences and conflicts among the cultures of the host countries, the differences and conflicts among the government policies of the host countries, the diversity and variety of the market conditions, and the diversity and variety of the innovative resources, all work on the internationalized technological innovation system of the enterprises, add to the environmental uncertainties, and force enterprises to make hard choices when deciding on their technological innovation strategies; as a result, enterprise innovation management is put in a situation teemed with a variety of conflicts and contradictions.

Under the new environmental conditions, the technological innovation strategy, organizational form and processes, and resource allocations of a business enterprise will all undergo fundamental changes. Considering strategy, breaking the traditional enterprise boundaries, expanding the field of view, allocating innovative resources based on their comparative advantages and environmental diversity, and choosing the right path of innovation have become the emphasis. In addition, to adapt to the demands of the new strategy and the need of the geographically dispersed R&D activities and to strengthen the communication and collaboration among those dispersed R&D institutions will become the guiding principles to construct a new system for internationalization of technological innovation. Global and multi-source acquisition, use, and integration of resources will become the main form of resource allocation under the new strategic framework.

This new mode of technological innovation needs to be matched with a new management model: taking the view of the company as a whole, functionally consolidating the geographically dispersed units, strengthening communication and cooperation among the various R&D units, and—at a higher level—maintaining the congruity and agreement among the business strategy, organizational form, and resource allocation scheme have become the focal points of efforts for preserving and improving the efficiency of technological innovation. The new management model of technological innovation will first ensure that the internationalized technological innovation system will be running efficiently; furthermore, it will help the business enterprise obtain various potential advantages of internationalization of technological innovation; and eventually, it will boost both the efficiency and effectiveness of enterprise technological innovation.

This study is based on the paradigm of internationalization of technology innovation (the SPRE model) sketched above; it will start with the idea of maintaining the congruity and agreement among the internationalization strategy, organizational form and processes, resource allocation, and environment support and go into the four detailed aspects of internationalization of technological innovation for Chinese business enterprises—strategy, organizational form, resource allocation, and environment.

With respect to the internationalization strategy of technological innovation, in our view, because of the complexity and unpredictability in the changes of the competitive environment, transforming the guiding strategic principle from one of purely market-oriented product development to the other based on improving enterprises' core competencies and capabilities is very important for technological innovation management. The strategic framework for internationalization of technological innovation based on building core competencies requires that Chinese business enterprises adopt a global vision and a long-term strategic perspective. The strategic goal is to enable the medium- to long-term business enterprises in China to globally acquire, allocate, and consolidate resources for technological innovation and to produce and operate at a highly internationalized level, that is, to enable them, first, to extend their innovative activities in the areas abroad, to diversify their sources of technological knowledge, and to obtain knowledge and technologies necessary for technological innovation from prestigious knowledge and technology centers worldwide, all through activities such as cross-border M&As, direct establishment of R&D institutions abroad, and the formation of international technology alliances; then, to upgrade core capabilities through internationally operated technological innovation activities; further, to promote the expansion of international production and other business activities; and, eventually, to achieve the globalization of their business operation as well as technological innovation.

On the basis of this understanding, we will study the characteristics and meaning of the internationalization strategy of technological innovation, relate this study to the investigation and analysis of the actual development in Chinese business enterprises, and propose—for these enterprises—a strategic framework, a strategic path of development, and an evolution model for internationalization of technological innovation, which is based on improving enterprises' core competencies.

With respect to the organizational form of internationalization of technological innovation, analysis will be made based on the strategic development path for internationalization of technological innovation for China. From the principle of agreement between strategy and organizational process, we conclude that the organizational form will undergo development from the star-shaped structure to the network structure and that the organizational process will change with the changes in the organizational structure. In this part of the work, we will perform a theoretical analysis on the organizational form and process design for internationalization of technological innovation, analyze the returned questionnaires of a survey, and, on the basis of the questionnaire analysis, propose an organizational form and process design framework that fit our internationalization strategy for technological innovation.

With respect to the resource allocation scheme for internationalization of technological innovation, we start from the principle of agreement among strategy, organizational process, and resource allocation and come to the conclusion that, during different development stages of internationalization of technological innovation, the acquisition, use, and integration of resources, such as information, personnel, and funding, should be dynamically adjusted with the changes in the strategy and the organizational form. The development from one with singular source of input to the other with multiple sources of input will be the direction of resource allocation. In this part of the work, we will perform a theoretical analysis on the resource allocation scheme, analyze the returned questionnaires of a survey, and, on the basis of the questionnaire analysis, propose a resource allocation scheme that fits our internationalization strategy and organizational form for technological innovation.

In our view, internationalization of enterprise technological innovation is a system's engineering endeavor. In a system of internationalization of technological innovation, the internal strategic measures, organizational forms, and resource allocation schemes together form the operating mechanism. Environmental factors are then the external support conditions for internationalized enterprise technological innovation. The efficiency of internationalization of technological innovation is a function of its environmental factors and the internal operating mechanism. Therefore, after analyzing the management of internationalization of technological innovation with respect to the strategy, organizational form, and resource allocation scheme, we will conduct a theoretical analysis on the support environment factors of internationalized enterprise technological innovation and, on the basis of this environmental factor analysis for the enterprises of China, present our policy proposal.

Finally, we will summarize the development model of internationalization of technological innovation for Chinese enterprises and put forward the evolutionary model and management emphasis for internationalization of technological innovation, which are based on building up technological capabilities for Chinese business enterprises.

1.2.2 Research methods

In this project, we first conducted a questionnaire survey and some business interviews. On the basis of our investigation on the present state of internationalization of technological innovation in Chinese business enterprises, we then adopted an approach with theoretical study combined with empirical study, qualitative study combined with quantitative study, and dynamic study combined with static study to conduct our research.

(1) Combining theoretical study with empirical study

On the one hand, we chose the latest theories on internationalization of technological innovation and on international economic competition and conducted theoretical analysis on the model and management of internationalization of technological innovation with respect to strategy, organizational form, resource allocation scheme, and environment support; on the other hand, we adopted methods such as questionnaire investigation, enterprise interviews, case studies on developments in the leading transnationals and domestic enterprises, and so on to engage in empirical study on the theories of internationalization of technological innovation; then, on the basis of both the theoretical and empirical studies, we proposed a theoretical model and a management approach for internationalization of technological innovation that are considered the best for Chinese business enterprises (see Figure 1-2).

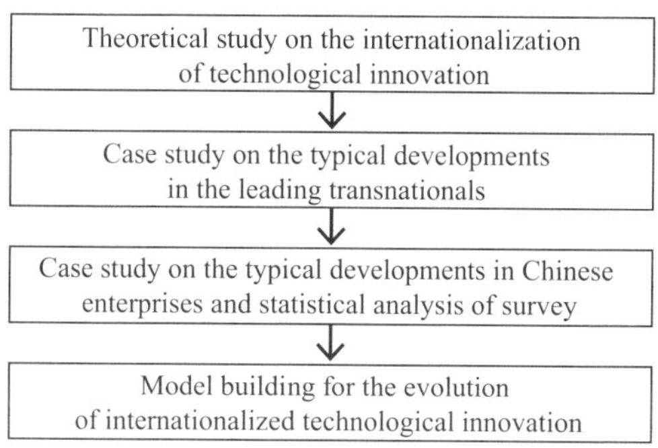

Figure 1-2 Technology route 1 of the research

(2) Combining qualitative study with quantitative study

When studying the general theoretical model for internationalization of technological innovation, we mainly laid emphasis on the qualitative analysis of the nature of the theory. When studying internationalization of technological innovation for Chinese business enterprises, we mainly relied on statistical analysis of the survey questionnaires and, after deriving quantitative facts, proposed a theoretical model for internationalization of technological innovation that fits the actual situations of these business enterprises. By quantitative study, we can ensure that our theoretical model is scientific and reliable.

(3) Combining dynamic study with static study

This project is not only to conduct analysis on the present state of the enterprise model of China but also, more importantly, to derive the path of evolution for internationalization of technological innovation of Chinese business enterprises, on the basis of the dynamic trend of internationalization of technological innovation for the leading transnationals and the dynamic trend of the participation in international competition by Chinese business enterprises. Thus, we combined dynamic study with static study, so that the findings of our research will meet the requirements of the changing environment (see Figure 1-3).

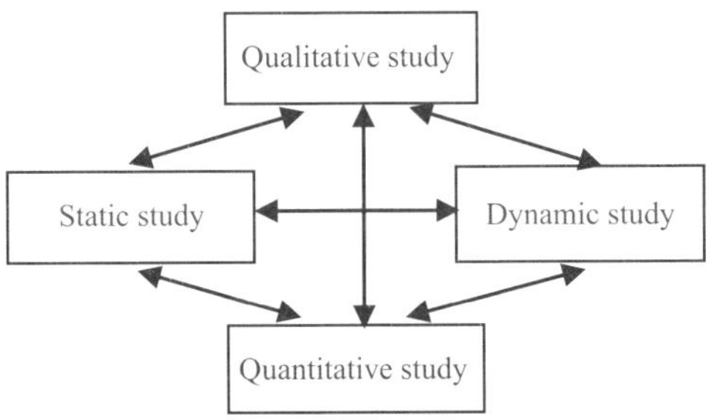

Figure 1-3 Technology route 2 of the research

References

Gerybadze, A. & Reger, G. Globalization of R&D: recent changes in the management of innovation in transnational corporations. *Research Policy*, 1999, 28(2-3): 251-274.

Lin, J. C. & Chai, Z. H. Analysis on the trend of internationalization of technological research and development in transnational corporations. *World Economy*, 1998(7): 35-39.

Niosi, J. & Godin, B. Canadian R&D abroad management practices. *Research Policy*, 1999, 28(2-3): 215-230.

Wang, X. C. Analysis of policies and directions of international cooperation in science and technology. *Science of Science and Management of S.&T.*, 1997(4): 27-29.

Yang, Z. G. & Bai, Z. G. On the national innovation system. *Journal of Shijiazhuang University of Economics*, 2000, 23(1): 29-31.

Chapter 2

Internationalization of Technological Innovation: A Literature Review

Foreign research on internationalization of technological innovation was basically developed alongside the gradual development of transnational corporations. In the emerging and early development stage of transnational corporations, there was not much research on internationalization of technological innovation and such research was done only in several aspects and within a small scale. From World War II to the present, internationalized production and trade have gone through unprecedented rapid development and internationalization of technological innovation by transnational corporations has reached a new historical stage; in addition to this, research on internationalization of technological innovation has entered a period of prosperity.

Presently, foreign research on internationalization of technological innovation mainly involves the following aspects: determining the factors of international R&D investment, the motivation for internationalization of technological innovation, the choice of location, the organizational form, the management approach, and so on.

2.1 Determinants of internationalization of technological innovation

Foreign scholars conducted studies in various ways to find the driving forces of internationalization of technological innovation, which mainly have the following three aspects.

(1) Supply of and demand for international direct investment

J. H. Dunning (1981)—a U.K. expert on transnationals—believes that internationalization by the transnationals with their technological innovation is an important part of their global competitive strategy. The ownership advantages and internationalization-specific advantages provide multinationals with the necessary conditions for direct investment in internationalization of technological innovation; then, location-specific advantages—which include the host country's resource endowments and environmental policies—provide them with sufficient conditions for such investment.

After analyzing the overseas R&D activities implemented by Swedish multinationals, Hakanson (1992) concluded that the overseas R&D activities are mainly demand-driven, for example, conducted to augment market potentials or to adapt to local market conditions. Geographic factors and market expansions are the determinants for internationalized R&D activities conducted by Swedish companies. Similarly, Pearce (1992) found that for both modifying existing products and developing new products, the demand factors are ranked first in the decision-making process for overseas R&D activities implemented by large-scale companies worldwide, while the supply factors (with respect to the host country's technological capabilities) are ranked second.

(2) Balancing the centrifugal/centripetal forces for internationalization of technological innovation in accordance with dichotomy (Terpstra, 1985; Pearce, 1989; Patel & Pavitt, 1991; Granstrand et al., 1992)

In a particular study, Granstrand et al. (1992) typically formulated the source model for the internationalization of corporate technological innovation at the corporate level, state level, and international level based on the division of the centrifugal/centripetal forces for internationalization of technological innovation.

Granstrand suggested that when viewed from the corporate level, the centrifugal force for internationalization—the promoting force—includes two aspects: the force of demand orientation and the force of supply orientation. The force of demand orientation includes transfer of replacement technology to subsidiary, integration of local production, subsidiary regional expansion, government control, and proximity to local customers and market. The force of supply orientation includes overseas M&A, proximity to overseas science and technology (S&T), proximity to low-cost/low-efficacy R&D human resources, and so on. The centripetal force of internationalization of technological innovation is the hindering force that includes the need to control R&D, concerns about leaking trade secrets, need to be close to the domestic market, economy of scale for R&D, cost of collaboration and exchange, historical inertia, and so on.

At the state level, the decision-making factors for internationalization of technological innovation mainly include culture of the company's home country and the host country, structure of the governing system, government policies, internal/external input/output market conditions, and so on.

At the international level, the decision-making factors include intercountry policies, agreement and system, integration of trade barriers, internationalism, global environmental factors, and so on.

Decisions about internationalization of technological innovation are made by an enterprise on the basis of the balance between the abovementioned two forces.

(3) Technological characteristics of the industry and national resource endowments

In a foreign research on internationalization of technological innovation, there is a traditional assumption: for large-scale companies from small industrialized countries (such as Belgium, the Netherlands, Sweden, and Switzerland), when formulating technological strategies, there is a strong tendency to internationalize, which is not so obvious for multinationals from large countries (Cantwell, 1995). In an industry, the less intensive the research is, the relatively more internationalized its technological activities are (Pavitt et al., 1989; Granstrand et al., 1997).

Cantwell and Kosmopoulou (2001) argued that there are deficiencies in such assumptions. They treated industry and technology as one aspect and investment abroad by a country and foreign investment in technological innovation in that country as another aspect and conducted studies on 792 of the world's largest industrial enterprises based on the data about patents granted by the United States to large-scale companies worldwide.

Cantwell and Kosmopoulou (2001) made two findings. First, the proportion of technological activities (via investment) abroad by an enterprise is closely related to the technological prowess of its industry group in each country and the proportion of technological activities by foreign enterprises (via investment in a host country) may probably be suppressed by the strength of the competitiveness of local enterprises. In turn, the degree of internationalization in technological development also depends on the level of interaction between local consumption company and local production company in industrial innovation (or corresponding national innovation systems). Second, when investing in innovative activities abroad, a leading business group in technology tends to develop supplementary and supporting technologies outside its own core area of expertise; development work in its core area will be concentrated in its home country. For example, a foreign company in a particular industry, on the one hand, may find that because the leading business group of the host country in the same industry, may be implementing a strategy to concentrate on technological development in its home country and thus may bring about fierce competition at home, it is impossible for them to gain a foothold there; however, powerful foreign companies in other industries, on the other hand, can develop technologies in that particular industry in that host country (to their own core areas of expertise, these technologies are not essential) to execute their diversification strategy.

Gerybadze and Reger (1999) have found through their research that rapid changes in technology and competition make it necessary for multinationals to have more than one center for R&D and innovation. They hoped that they can absorb knowledge from the outside (e.g., the center at home country absorbs knowledge from a certain region abroad) and that they can apply advanced technologies domestically. To absorb knowledge from the outside, multinationals need to set up institutions abroad in technologically advanced regions. In order to enhance the ability to absorb knowledge, multinationals have to establish stable and solid organizations in the S&T centers. Meanwhile, in order to respond to dynamic changes in regional technological advantages in a timely manner, it is a must to set up branch institutions in technologically advanced regions. In the areas of R&D, market, advanced manufacturing, and so on, multinationals must make necessary and intelligent choices; technology monitoring in different regions and building their own capabilities are complementary.

In addition, among different industries, the motivations for overseas R&D activities are different. A study by Dalton and Serapio (1999) also showed that in the electronics industry, the motivation for multinationals to conduct R&D activities in the United States is to acquire U.S. technologies in order to meet the needs of U.S. customers; in the biotechnology industry, the main objectives are technology acquisition, technology tracking, and making use of the good research environment in the United States. Considering the overseas R&D activities conducted by U.S. multinationals, Dalton and Serapio (1999) believed that market or demand is the main driving force for U.S. companies in the automotive industry, telecommunications, biomedical equipment, and industrial packaging to invest in overseas R&D activities and that technology and supply orientation are the main driving forces for companies in the computer software and hardware industry, chemical industry, and pharmaceutical industry to conduct overseas R&D activities.

2.2 Types of internationalization of technological innovation

The mission of internationalization of technological innovation is emphasized in the research studies done by foreign scholars. Many foreign scholars have conducted empirical studies on different sample sets (Jaffe, 1986; Cohen & Levinthal, 1989; Pearce, 1989, 1992, 1999; Hakanson & Nobel, 1993; Taggart, 1991, 1993; Casson & Singh, 1993; Kuemmerle, 1997, 1998, 1999a, 1999b; Zander, 1997, 1998, 1999; Cantwell & Janne, 1999; Dunning & Wymbs, 1999; Breschi et al., 1998; Cantwell & Piscitello, 1999, 2000) and have reached their conclusions.

In the early period, multinationals set up R&D institutions overseas to facilitate technology transfer. Mansfield et al. (1979) once conducted a very famous study on the R&D activities by U.S. companies, and the results showed that "the main purpose of a majority of the companies is to carry out technology transfer." Ronstadt

(1977, 1984) came to a similar conclusion: U.S. multinationals set up R&D units abroad to carry out technology transfer and to adapt their technologies to meet the requirements of the local markets. He also found that some of the technology transfer units have become indigenous technology units (ITUs) of technology transfer to develop new products for the local markets. These ITUs can also carry out more advanced technological activities, and become the technology center of the entire company (i.e., corporate technology unit, CTU) or even become the global technology center (i.e., global technology unit, GTU).

With rapid development of internationalization of technological innovation, overseas R&D activities are becoming ever more difficult to classify. Niosi and Godin (1999), by studying 16 large-scale and 2 small-scale Canadian companies, have found that commercial and pre-commercial R&D activities as well as product development activities are the most important in foreign subsidiaries. Other noteworthy activities include supporting the parent company's marketing efforts in the host country, providing the parent company with technical support, assisting the parent company in technology transfer, and so on. The secondary tasks of overseas research institutes are to provide technical support for external companies or organizations, to provide technical support for the government, to carry out basic research, and to conduct other studies. Technology transfer is no longer the main task for research institutes abroad that are owned by the multinationals.

By investigating and studying foreign R&D institutions in the United States and the overseas R&D activities by U.S. multinationals, Dalton and Serapio (1999) systematically summarized the classification of overseas R&D activities by U.S. multinationals and the R&D activities of foreign multinationals in the United States toward the end of the 20th century. They divided the foreign R&D activities of U.S. multinationals into two types: the market- or demand-driven type and the technology- or supply-oriented type. The market- or demand-driven type includes cases where the U.S. R&D activities are drawn to the host country for a number of factors: developing and designing products suitable for the local market; developing new products for the local market; and supporting the production, sales, and service units of the parent company in the host country. The technology- or supply-oriented type includes cases where the U.S. companies are pushed to carry out overseas R&D activities by a number of factors: mining R&D talent, developing new science and new technology, developing new products for global customers and monitoring technological development overseas, and participating in joint or cooperative research.

After conducting surveys on 32 multinationals and 238 research institutes, Kuemmerle (1997) proposed the concept of specialization: if an enterprise has established a number of R&D centers, it can make each laboratory specialize in an area of its strength. In the research, he classified the R&D institutions by their functions into two types, namely, home-based augmenting and home-based exploitation. Home-based augmenting institution emphasizes the ability of identifying and absorbing knowledge, obtaining knowledge from competitors and various universities and research institutes, and enabling ongoing information to flow from the laboratories abroad to the R&D center at the company's headquarters, while home-based exploitation institution emphasizes the ability of promoting internal transfer of knowledge, providing technical support for the company's production bases abroad, adjusting the degree of standardization to meet the local demands, and enabling ongoing information to flow from the R&D center at the company's headquarters to the laboratories abroad.

Pearce (1999) conducted an in-depth survey on 180 representative foreign R&D institutions in the United Kingdom and classified their functions into the following four categories: (i) supporting product manufacturing and process technologies, supporting manufacturing enterprises owned by multinationals, and operating in the United Kingdom; (ii) providing product manufacturing and process transformation, supporting manufacturing enterprises owned by multinationals, and operating outside the United Kingdom; (iii) cooperating and collaborating with other functions of the U.K. manufacturers and enhancing the ability of their products to better serve the target market; and (iv) specializing in basic research and applied studies, which are independent of localized manufacturing enterprises, under the control and coordination of the development plans and projects of the headquarters.

In addition, Hood and Young (1982, 1994), Hakanson and Nobel (1993), Chiesa (1996), and Taggart (1996) et al. also conducted research on this issue. According to their ideas, internationalization of technological innovation can be categorized into the following four types: (i) technology tracking, mainly for tracking changes in local technology and market; (ii) technical support and technical improvements, mainly for providing production and other business activities in branch units abroad with technical support; (iii) technology development, mainly for working with other functional parties in a subsidiary and focusing on the market conditions of the market to be serviced to carry out product development; and (iv) basic research and basic applied research, mainly for looking at the long-term developmental goals of the company and engaging in a precompetitive technology study to provide support for the long-term technological development of multinationals (see Table 2-1).

Some scholars conducted summary studies on the types of overseas R&D activities as a whole. Hewitt (1980) has found that internationalization of technological innovation can be divided into three categories: adaptational, market-oriented, and global. Similarly, Patel and Vega (1999) summarized the motivations for multinationals to internationalize technological innovation into three types: making adjustments to products, processes, and raw materials in order to adapt to the foreign market and providing technical support for overseas manufacturing units; tracking dynamic changes in overseas technological development; and developing core products and key technologies abroad.

The form and degree of internationalization of technological innovation change over time, thus resulting in varying research conclusions. By summarizing the current research conclusions, we categorize internationalization of technological innovation into three types: Type One is done to meet the special needs of the various customers worldwide and to adapt to local production conditions, in order to achieve technology localization, support the manufacturing enterprise of the company in the host country, and effectively configure enterprise capabilities; Type Two is done to set up a window of information, to track and obtain technologies from the host country and business competitors, to benefit from the spillover effect of local R&D activities, and to be the supplementary source of specialized technologies; Type Three is done to get close to knowledge centers, to make use of technology talent and research environment of the host country, to lower R&D cost, to promote the generation and diffusion of innovative activities, and to raise the overall technological capabilities.

Table 2-1 Types of overseas R&D institutions

<table>
<tr><th>Kuemmerle (1997)</th><th>Hood & Young (1982); Pearce (1989)</th><th>Ronstadt (1977)</th><th>Hakanson & Nobel (1993)</th><th>Von Boehmer (1995)</th><th>Chiesa (1996)</th></tr>
<tr><td>Home country—local enlarged laboratory</td><td></td><td></td><td>Technology monitoring</td><td>Focused on local R&D unit</td><td></td></tr>
<tr><td rowspan="4">Home country—local exploratory laboratory</td><td>Support laboratory</td><td>Technology transfer unit</td><td>Product support unit</td><td rowspan="2">Local problem-solving role</td><td>Product support unit</td></tr>
<tr><td>Local integrated laboratory</td><td>Local technical unit</td><td>Market support unit</td><td>Support laboratory</td></tr>
<tr><td></td><td>Corporate technical unit</td><td rowspan="2"></td><td>Cooperative researchers</td><td rowspan="2">Exploratory laboratory</td></tr>
<tr><td>Internationally interconnected laboratories</td><td>Global technical unit</td><td>Applied researchers</td></tr>
<tr><td></td><td></td><td></td><td>Multipurpose unit</td><td>Focused on global R&D unit</td><td>Test laboratory</td></tr>
<tr><td></td><td></td><td></td><td>Policy purpose unit</td><td></td><td></td></tr>
</table>

Source: Brockhoff, K. *Internationalization of Research and Development*. Berlin: Springer-Verlag, 1998.

2.3 Location choices of internationalization of technological innovation

Current empirical and normative studies conducted by foreign scholars on location choices for internationalization of technological innovation have concluded that the location choices are directly related to the overseas expansion of multinationals, motivations for internationalization, and the functions of overseas R&D institutions.

Mansfield et al. (1979), Lall (1979), Hirschey and Caves (1981), Pearce (1989), Zejan (1990), Reddy (1999), Hakanson (1992), and Kumar (1996) et al., at different times and from different perspectives, conducted studies on the factors affecting the location choices for internationalization of technological innovation related to overseas expansions by the multinationals. Their research showed that location choice for internationalization of technological innovation directly serves the expansion strategy of a multinational and that it is closely related to the degree of market penetration by the multinational in the host country, market size of the host country, local production orientation of the multinational in the host country, and the intensity of its technological activities, local R&D infrastructure, policy environment of the host country, and the level of local R&D activities conducted by competitors.

From this premise about the motivation for internationalization of technological innovation, Kumar (2001) assumed that three elements push for a location choice for overseas R&D activities—supporting overseas production, utilizing overseas lower cost R&D resources, and absorbing overseas R&D spillover—and studied the factors that affect the location choices for overseas R&D activities by U.S. and Japanese multinationals. The research found that location choice for R&D investment aimed at supporting overseas production is affected by the size of the local market because, by the economy of scale, adaptation to a very small-scale market is not cost-effective and that location choice for R&D investment aimed at utilizing lower cost R&D resources overseas is affected by the cost level of such human resources and by their availability: abundant R&D human resources make a country more attractive to R&D activities and a country with an easy access to lower cost R&D human resources will be successful in attracting overseas R&D investment. Location choice aimed at absorbing the spillover of overseas R&D unit is affected by the scale of the host country's technological endeavor—a country with large-scale technological activities has more opportunities to attract overseas R&D activities by multinationals.

Research by Cantwell and Hodson (1991), Hakanson (1992), Pearce and Singh (1992), and Florida and Kenney (1994) et al. showed that location choice for internationalization of technological innovation is directly related to the functions of overseas R&D unit. R&D activities in the production support or market-oriented category mainly serve to make adjustments to technologies and products from the home country and to adapt to the market conditions of the host country, and thus, proximity to existing manufacturing subsidiary of the parent company and to major clients is very important for location choices. The geographic distribution of R&D activities in the technology tracking or acquisition category is positively correlated with the geographic distribution of the overall innovative activities worldwide. The location choice for R&D activities in the resource utilization category is mainly affected by the supply of technological talent, facilities, and the degree of environmental assurance in the host country; R&D activities in the basic research-oriented category are done in cooperation with universities and research institutes in the host country, and thus, they need to be done near overseas technology centers of excellence. As a result, the location is usually somewhere close to famous universities or research bases of the same type.

According to a research by Reddy (1999), factors motivating multinationals to set up technology centers in developing countries include the following: (i) recruiting high-quality scientists and engineers, in order to alleviate the relative shortage of researchers in some fields as a result of the imbalance between supply and demand in industrialized countries; (ii) taking advantage of the differences in the salaries of researchers in developing countries and those in developed countries; and (iii) rationalizing business operations, that is, assigning the responsibility to certain subsidiaries to develop, manufacture, and sell certain products worldwide.

Generally, the higher the degree of market penetration by multinationals in the host country, the larger the market size, which means the production in the host country by multinationals is more oriented toward the regional market; the more intensive the technology, the more complete the local R&D infrastructure in a subsidiary. With an advantageous policy environment of the host country for the protection of intellectual property (IP), the entry of foreign R&D activities is more likely to happen.

Gerybadze and Reger (1999), by investigation, have found that qualitative factors and the integration of upstream and downstream activities increasingly affect the location choice for technological activities; for a multinational to choose a location for internationalization of technological innovation, considerations for implicit, knowledge-related, and value-adding learning and innovation factors far outweigh the considerations for quantity, cost, and efficiency. The considerations when choosing a location include the following: which location is in a region that is technologically advanced, renowned worldwide for research in a certain field, and has the advantages of conducting excellent research; which region has complex market and customer needs, which is conducive to innovation, and can increase the lead in a certain field enjoyed by the company; and which region has great potential for value appreciation. Furthermore, in which region, can manufacturing unit generate enough cash flow to support advanced research and to ensure the quality of research? Through researchers' effort and standardized network, which region can probably have an impact on system restriction and leading design and can help gain transfer advantage in the innovative activities worldwide? Which region has leading and innovation-oriented competitive companies and will the fierce competition there stimulate the development of new business and continued innovative activities? Therefore, multinationals would spend a lot of money searching for, evaluating, and selecting the best centers of excellence (there are usually only one or two such centers globally) and would strive to establish their R&D units in such centers of excellence.

In the internationalization process of multinationals, the choices of geographic locations reflected the abovementioned conclusions. A research by Gerpott (1994) indicated that, prior to the end of the 1970s, the general situation was that research and innovation centers were clustered in one particular country or region (technology centers were mainly concentrated in the United States; in some special fields, such as synthetic chemistry, they were mainly concentrated in Western

Europe). However, now the situation has changed because technological activities are becoming ever more concentrated in the United States, Japan, and Europe—a triangle as it were—and, at the same time, developing countries, especially newly industrialized nations, such as China, India, Israel, Singapore, Brazil, etc., are the emerging markets of technological activities by developed countries.

2.4 Development process and organizational form of internationalization of technological innovation

The development process of internationalization of technological innovation is closely related to that of the strategies for global competition and global innovation adopted by multinationals. Foreign scholars conducted research in this area at both micro and macro levels. At the micro level, when viewed from the perspective of corporations, there are mainly two forms—one with three stages and the other with five stages.

To explain the foreign R&D activities conducted by Swedish enterprises, Hakanson in 1990 proposed a three-stage development model. During the first stage, the R&D laboratory will be in one place; it is this one main unit that will develop all the innovative products; this unit will be supplemented by a small R&D laboratory network overseas, which will provide the technological support for technology transfer from the parent company to its subsidiaries. During the second stage, the polycenter stage, R&D laboratories will form a dispersed coalition; members of the group will carry out various endeavors, including overseas innovative research. The last and also the present stage is the stage of globalization, during which overseas R&D units, apart from carrying out the abovementioned activities, will enjoy more freedom in their research and the R&D laboratory at the parent company will perform greater communication and coordination functions.

Pearce and Papanastassiou (1999) also believed that when multinationals implement their strategies for global competition and global innovation, the development of their overseas R&D activities will go through the following three stages.

During Stage One, in order to make maximum use of the commercial potential of the current technology, multinationals must push their innovative products to the markets as quickly as possible; for this reason, there appear support-type overseas R&D institutions that are established to help overseas production units to make adjustments to their products and processes in order to adapt to the local production conditions and customer preferences.

During Stage Two, in order to maintain their competitiveness, multinationals must continually update their existing products and develop new products, and thus, there appear overseas R&D institutions that are integrated with the local subsidiaries. Such an R&D institution will mainly work with other functional units of a subsidiary and carry out product development activities based on the conditions of the market that the subsidiary services. The new ideas come either from its own applied study or from other R&D institutions of the company.

During Stage Three, multinationals, following their long-term development strategy, must continually renew their core technology, and thus, there appear internationally interconnected overseas R&D

institutions that aim at conducting basic or applied research. Such institutions are usually set up near the technology centers of excellence worldwide, and they collaborate with each other and form networked collaborative relationships and provide support for the long-term technology development of multinationals. Such overseas R&D institutions usually have very limited relationship with other functional units of the local subsidiary.

The situation of Japanese multinationals is very different from that of the Swedish ones; up until the 1980s, overseas R&D activities conducted by Japan had been making very slow progress and Japan lagged behind Western countries in the ratio of overall R&D expenditure. However, by the mid-1990s, the scale of Japan's overseas R&D activities had grown very fast, both in absolute and relative terms. Although the primary mission of Japanese overseas R&D institutions is product development and product adjustment, Japanese companies—compared with those in the West—place more emphasis on basic research. In order to compete in the Western markets, Japan must spawn its own disruptive innovation. It is very important for them to draw on Western sciences in order to maintain long-term competitiveness. Through a study on the internationalization by Japanese multinationals with their R&D activities, Sakakibara and Westney (1992) concluded that development of overseas R&D activities by Japanese companies has gone through the following five stages: (i) technology scan (small R&D institutions collecting overseas S&T information for use by the parent company in the home country); (ii) technology transfer; (iii) technological cooperation with local suppliers and consumers; (iv) local product development; and (v) long-term endeavors of basic research.

Presently, the three-stage theory of Hakanson is considered more universal.

At the macro level, the study is based on the evolutionary trend of the global innovation network. Bartlett and Ghoshal (1990), by dichotomy, summarized the evolutionary process of global innovative activities conducted by multinationals into three development stages, going from regional innovation organization serving local needs (local-for-local) to regional innovation center serving global needs (local-for-global) and further to global innovation center serving global needs (global-for-global) or global interlinked innovation network (globally linked).

Gerybadze and Reger (1999), from the knowledge management perspective, proposed a two-stage theory. In their view, when viewed as a whole, internationalized innovative activities have experienced two different paradigms. Innovative activities under the traditional paradigm can be viewed as a simple technological transformation process, that is, major R&D bases in the home country formulate new product concepts and new technological knowledge, and then such new concepts and technological knowledge are copied to the edges of the home country; this process can be understood as a one-way export process of technological knowledge. Being different from the traditional one, the new paradigm has the following characteristics: (i) there are multiple knowledge sources set in multiple geographic locations; (ii) there are both learning imports and exports, which are different from one-way information flow; and (iii) interactions by technology transfer occur between different internal organizational units and also between different geographic regions. This new paradigm for internationalization of technological innovation is built on the organizational form of a regional or global innovation network, and its main objective is to create technological knowledge across a wider area.

The organizational form of internationalization of technological innovation is consistent with its organizational function, and it changes with the development process of internationalization. Brockhoff (1998), on the basis of the organizational functions, classified various organizational structures for internationalization of technology innovation into three types: star or center-edge, multiple regional capability centers, and globally interconnected research organizational structure. In an empirical study, Bartlett and Ghoshal (1990) concluded that there are four types of universal overseas innovative organizations: global innovation center, local innovation center serving local needs, local innovation center serving the global market, and global interlinked innovation network. Gassmann and Von Zedtwitz (1999) divided the organizational forms for internationalized R&D networks into five types: ethnocentric centralized R&D network, geocentric centralized R&D network, polycentric decentralized R&D network, R&D hub model, and integrated R&D network. Other scholars (Chiesa, 1996; Asakawa, 2001) have also drawn similar conclusions.

The organizational forms for internationalization of technological innovation are not set in stone; to the contrary, they are changing constantly. To improve the efficiency of global R&D activities, many companies have gradually integrated their R&D activities into a small number of technology centers of excellence. To meet the needs of international market, R&D organizations have become ever more open, R&D processes have got closer to the international market and knowledge centers, and the organizational forms have been moving toward the direction of integrated R&D network (Von Zedtwitz, 2002).

2.5 Management of internationalization of technological innovation

Existing research mainly focuses on the trends of development and the determinants of internationalization of technological innovation. Some scholars conducted studies on the relationship between corporate technology strategies and competitive advantages, but such studies are mainly limited within the context of one country and are rarely done on ways to manage dispersed R&D activities as a whole (Ensign, 2000). Thus, Bartlett and Ghoshal in 1988 noted that "out of 4000 studies on innovation, there is not one with a real focus on the processes of internationalized innovation by multinationals." There has not been much systematic research about how to manage relatively dispersed R&D activities following the trend of internationalization. Now, with the deepening of internationalization of technological innovation activities, scholars have begun to pay attention to the question of how to manage geographically dispersed R&D activities as a whole.

With an increase in the geographically dispersed R&D activities, management complexity has become an urgent issue in regard to internationalization of technological innovation. Since R&D branch units and companies' headquarters often have different understandings, hold different opinions about information sharing, and have disagreements about control and autonomy, there are many contradictions within a global innovation network (Asakawa, 2001). The problem of a globalized R&D unit is not a structural one. Companies need to balance cost, speed, and quality and thus are often faced with the dilemmas of centralization and decentralization, creativity and norm, process and level, face-to-face communication and communication through telecommunication equipment, and so on (Von Zedtwitz & Gassmann, 2002).

Centralization and decentralization, integration, communication, and human resources have become the focus for management of global innovation network.

2.5.1 Centralization and decentralization

When formulating strategic decisions, a company will be faced with this question: how much autonomy should be granted to local R&D institutions? Too much freedom can inadvertently lead to waste and difficulty in control. Excessive control will stifle creativity and prevent the full utilization of local capabilities (Von Zedtwitz, 2002). Widespread use of the Internet has removed the geographic barriers to information. Now, even a very small company can search all over the world for the knowledge it needs. However, this method is mainly used to find data or information with a relatively high level of structure—explicit knowledge. R&D experience and tacit knowledge cannot be downloaded from the Internet; tacit knowledge resides in persons. To successfully transfer such knowledge requires direct personal contact; thus, dispersed R&D activities will continue to exist in the foreseeable future.

Based on the level of technological autonomy, Niosi and Godin (1999) divided overseas technological activities conducted by Canadian multinationals into the following three categories. (i) Related diversification—parent company grants the subsidiary a lot of freedom: some locals assume management positions; funding is provided by subsidiaries; and project decisions can be made locally based on the local market conditions. In this way, the coordination and communication between R&D subsidiary and the parent company will no longer pose a serious problem. (ii) Vertical integration—some of the R&D managers in the subsidiaries are Canadians; the creative ideas for some projects come from the Canadian parent company, and the funds are also partially provided by the parent company. There are some communication and coordination problems between R&D units and subsidiaries; this type of international network is different from any of those put forward by other studies. (iii) Global strategy (equivalent to GTU)—parent companies usually provide a large amount of research funding and participate in project decision-making. Among senior management of overseas R&D units, the number of Canadians significantly increases. The issues arising from such global networks built up by multinationals have been dealt with in previous papers (Hakanson, 1990; Chiesa, 1996). The autonomy of overseas R&D institutions depends on the allocation of activities within the multinational—R&D institutions of the related diversification type enjoy a higher level of autonomy, those of the vertical integration type require more coordination and technology transfer, and there are more restrictions in a global network and there needs to be more coordination and more international flow of talent, funding, technology, and creative ideas from the parent company to the established R&D institutions.

2.5.2 Integration of internationalized R&D activities

Business enterprises have been continuously working to integrate the chain of innovation process. Past questionable practice to isolate basic research, applied research, development, product, and practical needs should be corrected. Multinationals and leading research centers continue to improve their competitiveness by virtue of the interactions between research, advanced manufacturing, and advanced market. The integration of product development process and related technology and the cross-border integration of R&D, marketing, production, and product have become key issues in innovation management (Leonard-Barton, 1995; Nonaka & Takeuchi, 1995).

Gerybadze and Reger (1999) believed that it is the basic type of globalized innovation that determines the choice of integration mechanism for a transnational corporation (see Figure 2-1). In Type-A and Type-C innovations, the choice of organizational design and cooperative mechanism appears to be more direct. Among innovations based on science and research, leading R&D centers are always closely linked to the center of the company and they always implement centralized decision-making. Research-centered corporate culture promotes continued innovation. Type-C innovations have a more pressing need for cross-function integration, and they require a systemic center running throughout the value chain. Management capabilities for many of the most important functions and system integration are set up in an important country.

In possession of significant assets in the home country	A · A close combination of leading research centers and company's headquarters; · Centralized decision-making in the home country; · Technology platform based in the home country; · An enterprise culture with uplifting research mainly done in the home country	C · Integration between R&D and leading market mainly done in the home country; · Cross-function teams in the territory of one country; · Internal markets and contracts among business units and functions
In possession of significant assets abroad	B · Research projects involving multiple R&D laboratories; · Global team management; · Effective combination of R&D and technology strategy; · A strong culture of cooperation	D · Establishing effective business units and branch institutions in leading markets; · Cross-function integration in leading markets; · Ensuring that company's headquarters fully understand business developments
	Innovation based on science and research	Innovation solidly based on leading markets

Figure 2-1 Integration mechanism determined by the basic type of innovation

Source: Gerybadze, A. & Reger, G. Globalization of R&D: recent changes in the management of innovation in transnational corporations. *Research Policy*, 1999, 28(2-3): 251-274.

For Type-B innovation, the company needs to obtain important resources globally from various places. They organize R&D activities in the same way as organizing a network with important laboratories located in different places, and thus, an effective global management team is necessary. Team members from different cultural and educational backgrounds need to be integrated, which can be achieved through conducting effective joint research tasks and implementing technological strategies. However, contrary to the optimistic descriptions made by Bartlett and Ghoshal (1990) and Hedlund (1990), Gerybadze and Reger (1999) believed that in most cases, a company integrates only a few of its leading laboratories.

Finally, Type-D innovation strategy makes it necessary for the company to cross the boundaries of organizational entities (functional and operational units) and to cross the national borders to manage innovation processes. On the one hand, somewhat tighter control mechanism is needed to reduce the complexity, and on the other hand, it is necessary to adhere to the polycenter structure for customer knowledge. Their approach is to link these centers to the overall center as much as possible. With the company's business staff and key members of teams posted abroad, the critical element for the company's success lies in whether or not the headquarters of the company can fully understand the dynamic business developments and whether or not they can shift key resources to such businesses.

Von Zedtwitz (2002) has found that the R&D process can be divided into the following three phases to expedite the progress of innovation. (i) Early phase—in the early phase of R&D, the flow is nonlinear and good ideas must be formulated. There are no geographic limitations as to where to look for knowledge of new technologies; however, inventions and innovations are usually regional activities. System architecture needs to be clearly defined. (ii) Project implementation—project development must be strictly and efficiently implemented. During this phase, global team cooperation is possible, but responsibilities and interfaces need to be clearly defined. (iii) Market introduction—there may be different strategies for this phase: introduction to market can happen in one market at a time or new product introduction can be coordinated and can happen in all the markets at the same time.

2.5.3 Exchange and communication

Communication technologies can affect the content and frequency of information exchange within an organization (Brockhoff, 1998). Communication is the cornerstone for the success of an international business; however, there are not many scientific studies in this area.

Studies on communication within an international innovation team among members from different companies, different countries, and different cultural backgrounds are absolutely necessary (Moenaert et al., 2000).

Internationalized R&D activities connect the globally dispersed R&D institutions into the company's innovation network, and now, geographic distance becomes a significant impediment to communication; thus, effective networked operation requires the support of complex global information technology infrastructure. Unprecedented information resources and information technologies have provided new opportunities for the management of R&D. E-mails, shared databases, and remote login technology can greatly improve the efficiency of cross-border R&D processes. Without modern information and communication technologies, there would not have been cross-border R&D projects. However, for communication in the processes of internationalized R&D activities, information technology is the only necessary means (Von Zedtwitz, 2002). Information technology can promote the exchange of explicit knowledge; the exchange of tacit knowledge can only be achieved with face-to-face communication. Without face-to-face communication, it would not be possible to form the necessary trust among members of a project and, thus, it would be more difficult to break cultural barriers; especially early in the team spirit building phase of a project, trust in project leaders and interface definition of the system structure are very important. Although latest information technology has already been experimenting with the transmission of tactile and sensory information, there may not be remote communication technology in the next ten years, which will be able to completely replace face-to-face communication.

In order to promote effective communication in R&D processes, Moenaert et al. (2000) conducted systematic studies on communication in the processes of internationalized innovation. In their view, an innovation team is in itself an information processing system while an organizational structure capable of successfully managing an internationalized innovation project is a structure that can match information processing capabilities with information processing needs. Therefore, there are five necessary conditions to engage in communication in internationalized innovation: transparency of the communication network, knowledge codification, knowledge reliability, communication cost, and privacy. Moenaert et al. (2000) found that in order to achieve effective communication, organizations must develop capabilities at the enterprise level and at the team level.

At the R&D staff level, scientists are somewhat arbitrary and product development engineers are very strict; then, R&D staff members and members of other functional units oftentimes have different understandings about a project, and there are oftentimes friction among such members. It is necessary to implement frequent job rotations among different departments so that the members of such departments can achieve better understanding among themselves. Meanwhile, engineers and researchers should be given plenty of freedom, so that they will form an unofficial communication network and promote the sharing of tacit knowledge (Von Zedtwitz, 2002).

2.5.4 Human resource management

The core of every innovation is a thinking created by man. To ensure effective communication and coordination within a geographically dispersed internationalized innovation network and to maintain its creativity and high efficiency, effective human resource management is one of the keys.

The selection of team leaders is very important to the success of such innovation; the management and the team first need to select and train project managers; especially, during the phase for concept formation, R&D managers are the key persons to form consistent goals and define a clear-cut system structure. Moenaert et al. (2000) and Von

Zedtwitz and Gassmann (2002) believed that good leaders should be able to process cross-function and cross-subject information; should be able to maintain the continuity of the project, select necessary capabilities, bridge cultural differences, control information flows, and harmonize the needs of different individuals; and should be able to correctly make use of the knowledge within a multinational about how to operate the entire product development process and to solve the problem of low efficiency, which exists internally within a multinational. If inexperienced project leaders or partners are chosen, they may not only hinder the project but also often lead the project into difficult situations from the very beginning. Because the criterion used by many companies to choose project managers is "the most easily found, not the most suitable," the appointed project managers are often lacking in knowledge and experience and thus are not able to lead the project process. They also do not know who, within the communication network of the multinational, are the potential partners, and thus, they would fail to consult these people and such failure will lead to a chaotic atmosphere within the organization.

The need for effective communication and coordination and for maintaining creativity and efficiency of the organization also places requirements for employees, as they must be able to use modern communication technology, have good communication skills, be adaptable, be willing to learn, be able to work within a team, and be able to do creative work.

As mentioned above, obtaining low-cost R&D human resources and tapping into overseas R&D talent pool is one of the motivations for internationalization, and thus, personnel selection for R&D is often done around the world.

2.6 Chapter summary

Studies by foreign scholars have covered aspects, such as motivation, strategies, organizational forms, and management, of internationalization of technological innovation; such studies have basically been built on the theoretical framework for internationalization of technological innovation. However, there are still some problems in such research done by the foreign scholars.

First, the objects of study are mainly multinationals and the research mainly aims at the problem of internationalization by business enterprises in developed countries with their technological innovation; research targeted at enterprises in developing countries is still almost non-existent. Compared with transnationals, enterprises in developing countries operate entirely at different levels of strengths and are faced with very different competitive environments, and hence, their paths toward internationalization of technological innovation will necessarily be very different from those of transnationals. Research on internationalization by business enterprises in developing countries with their technological innovation is an important aspect of the studies on internationalization of technological innovation. Without a theory on internationalization of technological innovation for enterprises of developing countries, the system of theories in this area is incomplete.

Second, studies are mainly done at the industry level or with regard to the broader economy; those that are focused on individual enterprises, especially the studies of the management personnel of such enterprises, are obviously lacking. From the perspective of business management, studies on the management of geographically dispersed global technological innovation system should be the focus of future research.

Third, studies are mainly done from the perspective of the traditional paradigm for the management of multinationals; such traditional paradigm ignores the flexibility and dynamic nature of new organizations and systems. Research with a focus on the new trend of global innovation, that is, on internationalization of technological innovation with an emphasis on the creation of knowledge and capabilities, is lacking.

Fourth, systematic studies on internationalization of technological innovation that is based on the idea of maintaining the congruity and agreement among internationalization strategy, organizational form and processes, resource allocation, and environment support are relatively lacking.

For Chinese enterprises, because of limited business strengths, low levels of internationalized production and operation, and low levels of technological innovation management, internationalization of technological innovation either has not yet started or is still in the initial stage. The tasks of China's theoretical circle in the field of technological innovation include the following: (i) carefully analyzing the internationalization by leading multinationals of their technological innovation; (ii) drawing on foreign research on internationalization of technological innovation; (iii) developing—with a focus on the actual situation after China's accession to the WTO—the theory, operational mode, and management method of internationalization of technological innovation that fit the situation of China; (iv) providing guidance for Chinese business enterprises to conduct internationalization of technological innovation and fully participate in international competition, and developing new theories on the basis of past research.

References

Asakawa, K. Organizational tension in international R&D management: the case of Japanese firms. *Research Policy*, 2001, 30(5): 735-757.

Bartlett, C. A. & Ghoshal, S. Managing innovation in the transnational corporation. In: Bartlett, C. A., Doz, Y. & Hedlund, G. (eds.). *Managing the Global Firm*. London: Routledge, 1990.

Breschi, S., Lissoni, F. & Malerba, F. Knowledge proximity and firm's technological diversification. Milan: CESPRI, Bocconi University, 1998.

Brockhoff, K. *Internationalization of Research and Development*. Berlin: Springer-Verlag, 1998.

Buckley, P. J. & Casson, M. *Multinational Enterprises in the World Economy*. London: Elgar, 1992.

Cantwell, J. A. The globalization of technology: what remains of the product cycle model?. *Cambridge Journal of Economics*, 1995, 19(1): 155-174.

Cantwell, J. A. & Hodson, C. Global R&D and UK competitiveness. In: Casson, M. (ed.). *Global Research Strategy and International Competitiveness*. Oxford: Basil Blackwell, 1991.

Cantwell, J. A. & Janne, O. E. M. Technological globalization and innovative centers: the role of corporate technological leadership and locational hierarchy. *Research Policy*, 1999, 28(2-3): 119-144.

Cantwell, J. A. & Kosmopoulou, E. Determinants of internationalization of corporate technology (DRUID working papers No 01-08). Copenhagen: DRUID, 2001.

Cantwell, J. A. & Piscitello, L. The emergence of corporate international networks for accumulation of dispersed technological competence. *Management International Review*, 1999(39)(Special Issue): 123-147.

Cantwell, J. A. & Piscitello, L. Accumulating technological competence: its changing impact on corporate diversification and internationalization. *Industrial and Corporate Change*, 2000, 9(1): 21-51.

Casson, M. & Singh, S. Corporate research and development strategies: the influence of firm, industry and country factors on the decentralization of R&D. *R&D Management*, 1993, 23(2): 91-107.

Chiesa, V. Managing the internationalization of R&D activities. *IEEE Transactions on Engineering Management*, 1996, 43(1): 7-23.

Cohen, W. & Levinthal, D. Innovation and learning: the two faces of R&D. *The Economic Journal*, 1989, 99(397): 569-596.

Cohen, W. & Levinthal, D. Absorptive capacity: a new perspective on learning and innovation. *Administrative Science Quarterly*, 1990, 35(1): 128-152.

Dalton, D. H. & Serapio, M. G. *Globalizing Industrial Research and Development*. Washington, DC: U.S. Department of Commerce, Technology Administration Office of Technology Policy, 1999.

Dunning, J. H. *Multinationals, Technology and Competitiveness*. London: Unwin, 1988.

Dunning, J. H. Multinational enterprise and the globalization of inventory capacity. In: Granstrand, O., Hakanson, L. & Sjolander, S. (eds.) *Technology Management and International Business*. Hoboken, NJ: Wiley, 1992.

Dunning, J. H. & Wymbs, C. The geographical sourcing of technology-based assets by multinational enterprises. In: Archibugi, D., Howells, J. & Michie, J. (eds.). *Innovation Policy in a Global Economy*. Cambridge: Cambridge University Press, 1999.

Ensign, P. C. Innovation in the multinational firm with globally dispersed R&D: technological knowledge utilization and accumulation. *The Journal of High Technology Management Research*, 2000, 10(2): 203-221.

Florida, R. & Kenney, M. The globalization of Japanese R&D: the geography of Japanese R&D investment in the United States. *Economics Geography*, 1994, 70(3): 344-369.

Gassmann, O. & Von Zedtwitz, M. New concepts and trends in international R&D organization. *Research Policy*, 1999, 28(2-3): 231-250.

Gerpott, T. Successful integration of R&D functions after acquisitions: an exploratory empirical study. *R&D Management*, 1994,

25(2): 161-178.

Gerybadze, A. & Reger, G. Globalization of R&D: recent changes in the management of innovation in transnational corporations. *Research Policy*, 1999, 28(2-3): 251-274.

Granstrand, O., Hakanson, L. & Sjolander, S. *Technology Management and International Business: Internationalization of R&D and Technology*. Chichester: Wiley, 1992.

Granstrand, O., Patel, P. & Pavitt, K. L. R. Multi-technology corporations: why they have "distributed" rather than "distinctive core" competencies. *California Management Review*, 1997, 39(4): 8-25.

Hakanson, L. International decentralization of R&D: the organizational challenges. In: Bartlett, C. A., Doz, Y. & Hedlund, G. (eds.). *Managing the Global Firm*. London: Routledge, 1990.

Hakanson, L. Locational determinants of foreign R&D in Swedish multinationals. In: Granstrand, O., Hakanson, L. & Skolander, S. (eds.). *Technology Management and International Business*. Chichester: Wiley, 1992.

Hakanson, L. & Nobel, R. Determinants of foreign R&D in Swedish multinationals. *Research Policy*, 1993, 22(5-6): 369-411.

Hewitt, G. Research and development performed abroad by US manufacturing multinationals. *Kyklos*, 1980, 33(2): 308-327.

Hirschey, R. C. & Caves, R. E. Research and transfer of technology by multinational enterprises. *Oxford Bulletin of Economics and Statistics*, 1981, 43(2): 115-130.

Hood, N. & Young, S. US Multinational R&D: corporate strategies and policy implications for the UK. *Multinational Business*, 1982(2): 10-23.

Hood, N., Young, S. & Lal, D. Strategic evolution within Japanese manufacturing plants in Europe: UK evidence. *International Business Review*, 1994, 3(2): 97-122.

Jaffe, A. B. Technological opportunity and spillovers of R&D: evidence from firms' patents, profit and market value. *American Economic Review*, 1986, 76(5): 984-1001.

Kuemmerle, W. Building effective R&D capabilities abroad. *Harvard Business Review*, 1997, 75(2) 61-67.

Kuemmerle, W. Strategic interaction, knowledge sourcing and

knowledge exploitation in foreign environments: an analysis of foreign direct investment in R&D by multinational companies. In: Hitt, M. A., Richard, J. E. & Nixon, R. D. (eds). *Managing Strategically in an Interconnected World.* Chichester: Wiley, 1998.

Kuemmerle, W. The drivers of foreign direct investment into research and development: an empirical investigation. *Journal of International Business Studies*, 1999a, 30(1): 1-24.

Kuemmerle, W. Foreign direct investment in industrial research in the pharmaceutical and electronics industry—results from a survey of multinational firms. *Research Policy*, 1999b, 28(2-3): 179-193.

Kumar, N. Intellectual property protection and location of overseas R&D activities by multinational enterprises. *World Development*, 1996, 24(4): 673-688.

Kumar, N. Determinants of location of overseas R&D activity of multinational enterprises: the case of US and Japanese corporations. *Research Policy*, 2001, 30(1): 159-174.

Lall, S. The international allocation of research activity by US multinationals. *Oxford Bulletin of Economics and Statistics*, 1979, 41(4): 313-331.

Leonard-Barton, D. *Wellsprings of Knowledge: Building and Sustaining the Sources on Innovation*. Boston: Harvard Business School Press, 1995.

Mansfield, E., Teece, D. J. & Romeo, A. Overseas research and development by US-based firms. *Economica*, 1979, 46(182): 187-196.

Moenaert, R. K., Caeldries, F., Lievens, A. & Wauters, E. Communication flows in international product innovation teams. *Journal of Product Innovation Management*, 2000, 17(5): 360-377.

Niosi, J. & Godin, B. Canadian R&D abroad management practices. *Research Policy*, 1999, 28(2-3): 215-230.

Nonaka, I. & Takeuchi, H. *The Knowledge Creating Company: How Japanese Companies Create the Dynamics of Innovation*. New York: Oxford University Press, 1995.

Patel, R. D. & Pavitt, K. Large firms in the production of world's technology: an important case of non-globalization. *Journal of*

International Business Study, 1991, 22(1): 1-22.

Patel, P. & Vega, M. Patterns of internationalization of corporate technology: location vs. home country advantages. *Research Policy*, 1999, 28(2-3):145-155.

Pavitt, K., Robson M. & Townsend, J. Technological accumulation, diversification and organization in UK company, 1945–1983. *Management Science*, 1989, 35(1): 81-99.

Pearce, R. D. *The Internationalization of Research and Development by Multinational Enterprises*. London: Macmillan, 1989.

Pearce, R. D. Factors influencing the internationalization of R&D in multinational enterprises. In: Buckley, P. J. & Casson, M. (eds.). *Multinational Enterprises in the World Economy*. London: Elgar, 1992.

Pearce, R. D. Decentralized R&D and strategic competitiveness: globalized approaches to generation and use of technology in multinational enterprises (MNEs). *Research Policy*, 1999, 28(2): 157-178.

Pearce, R. D. & Papanastassiou, M. Overseas R&D and the strategic evolution of MNEs: evidence from laboratories in the UK. *Research Policy*, 1999, 28(1): 23-41.

Pearce, R. D. & Singh, S. *Globalizing Research and Development*. London: Macmillan, 1992.

Reddy, P. *Globalization of Corporate R&D*. London: Routledge, 1999.

Ronstadt, R. C. *R&D Abroad by US Multinationals*. New York: Praeger, 1977.

Ronstadt, R. C. R&D abroad by US multinationals. In: Stobaugh, R., Wells, Jr., L. T. (eds.). *Technology Crossing Borders*. Boston: Harvard Business School Press, 1984.

Sakakibara, K. & Westney, D. E. Japan's management of technological innovation. In: Rosenberg, N., Landau, R. & Mowery, D. C. (eds.). *Technology and the Wealth of Nations*. Stanford: Stanford University Press, 1992.

Terpstra, V. International product policy: the role of foreign R&D. In: Wortzel, H. V. & Wortzel, L. V. (eds.). *Strategic Management of Multinational Corporation: The Essentials*. New York: Wiley, 1985.

Taggart, J. H. Determinants of foreign R&D locational decisions in the pharmaceutical industry. *R&D Management*, 1991, 21(3): 229-240.

Taggart, J. H. Strategy conflict in the MNE: parents and subsidiary. Paper presented to 19th annual conference of European International Business Association, 1993.

Taggart, J. H. Multinational manufacturing subsidiaries in Scotland: strategic role and economic impact. *International Business Review*, 1996, 5(5): 447-468.

Von Zedtwitz, M. Organizing global R&D: challenges, dilemmas and future trends. Miami: IAMOT2002, The 11th International Conference on Management of Technology, 2002.

Von Zedtwitz, M. & Gassmann, O. Market versus technology drive in R&D internationalization: four different patterns of managing research and development. *Research Policy*, 2002, 31(4): 569-588.

Young, S., Hood, N. & Dunlop, S. Global strategies, multinational subsidiary roles and economic impact in Scotland. *Region Studies*, 1988(22): 487-497.

Zander, I. Technological diversification in the multinational corporation—historical evolution and future prospects. *Research Policy*, 1997, 26(2): 209-227.

Zander, I. The evolution of technological capabilities in the multinational corporation—dispersion, duplication and potential advantages from multinationality. *Research Policy*, 1998, 27(1): 17-35.

Zander, I. How do you mean "global"? An empirical investigation of innovation network in the multinational corporation. *Research Policy*, 1999, 28(2-3): 195-213.

Zejan, M. C. R&D activities in affiliates of Swedish multinational enterprise. *Scandinavian Journal of Economics*, 1990, 92(3): 487-500.

Chapter 3

Internationalization of Technological Innovation: Strategic Framework and Strategic Approach

3.1 Theories of technological innovation and technological innovation strategy—origin of internationalization strategy of enterprise technological innovation

The theory of innovation was first proposed by Schumpeter in 1912 in his book *The Theory of Economic Development*, which marked the beginning of study on the role of technology in economic growth.[1] Schumpeter not only regards innovation as the core concept to the understanding of the capitalist system and its development but also believes that innovation is a spontaneous,

[1] Schumpeter said this about innovation: "This concept includes the following five situations: (i) adopting a new product—a product that is not yet familiar to consumers—or a new feature to a product; (ii) adopting a new production method—a method that is not yet proven by experience in production department, which never needs to be based on new findings in science and which can also be a new way to process a product commercially; (iii) opening up a new market—a market that a certain manufacturing sector of a country has never entered, regardless of whether or not that market existed previously; (iv) plundering or controlling a new supply source of raw materials or intermediate products, regardless of whether the source existed before or it came into existence for the first time; (v) achieving reorganization of any industry, such as creating a monopoly (for example, through 'trust' formation) or breaking a monopoly."

qualitative, and revolutionary phenomenon rather than a quantitative one. Nikolai Kondratiev during the 1930s proposed the long wave theory, which from another angle highlights the role of technology in economic long wave, or long-term cyclical growth of the economy. After Schumpeter, especially since the middle of the 20th century, economists in technological innovation have come to the belief that technological innovation is the principal factor of economic growth.[1]

If Schumpeter's ideas mainly illustrate the importance of innovation with regard to macroeconomy, then the results of later research on technological innovation based on Schumpeter's ideas have greater significance in guiding mesoeconomy at the industry level and microeconomy entities, i.e., enterprises.

Freeman and Phillips separately drew two innovation models from

[1] Solo made pioneering contributions in explaining the effect of technological progress on economic growth. In his 1957 paper, he creatively put forward a growth model that involves technological progress and thereby solved the puzzle that economic growth cannot be fully explained by capital and labor. In this model, Solo modified the then-existing production function, added a technology factor *A*, and thus changed the description of the function to *AF (K, L)*. To simplify the model, he made the assumption that technological progress is Harrod-neutral, $\frac{\Delta Y}{Y} = (1-k_1)\cdot\frac{\Delta L}{L} + (1-k_2)\cdot\frac{\Delta K}{K} + \frac{\Delta A}{A}$, and from there drew the following conclusion: economic growth is driven by technological advancement, labor, and capital. He used this model and estimated that, between 1909 and 1949, 80% of the growth in the U.S. economy was driven by technological advancement; from there, he concluded that technological advancement plays a vital role in economic growth.

Schumpeter's theory.[1] In 1966, Schmukler proposed the demand-pull theory for the driving mechanism of S&T industrialization.[2] Schmukler's theory aroused heated debate in theoretical circle and led to more in-depth empirical studies and theoretical analyses of technological innovation. At the end of the debate, Rosenberg (1982) pointed out that S&T—as a basic, evolving knowledge base—and market demand work together interactively and play a central role in innovation and that ignoring either of the two will lead to erroneous conclusion and policy; hence, he merged the technology-push theory and the demand-pull theory into an organic whole. Roy and Cross in 1983 proposed a model and pointed out that "the conception of technological ideas is at the center of technological innovation that is based on the combination of basic research and market research, and the main path of innovation is the process starting from the core concept and going through invention, prototype development, manufacturing development, production and sales until the innovation is fulfilled." Kline and Rosenberg proposed a further conclusion by their research and believed that innovation is a complex interactive process, and

[1] Freeman summed up Schumpeter's Model I into the following: "(i) From the latest advances in science to important inventions exist non-specific flows; to existing enterprises and market structure, and to measurable market demands, these flows are exogenous. (ii) A group of entrepreneurs recognize the future potential of such innovations, and are willing to take the risk of development and innovation. (iii) Once a major innovation has happened, it will lead to imbalance in the existing market structure, and successful innovators will gain extra growth rate and temporary monopolistic profits. However, such monopoly will be weakened by a large number of second generation innovators entering the market, and thus bring about the phenomenon of periodic clustering." Phillips then pointed out that there is also a modification to the abovementioned model called Schumpeter's Model II. In Model II, Schumpeter replaced the role of entrepreneurs with large enterprises and pointed out that R&D activities are mainly conducted by the internal R&D units to business enterprises and that exogenous S&T and endogenous S&T develop together as companions.

[2] Strictly speaking, Schmukler himself does not believe that "the force of demand is the only factor in inventive and innovative activities"; he was only trying to rectify the extreme imbalance of the innovation model. However, Schmukler in his writings focuses on needs analysis and thus is considered the pioneer of the demand-pull theory.

based on this belief, they also proposed a chain-linked model.[1] The abovementioned conclusions from the research studies on technological innovation are significant in guiding enterprises in conducting technological innovation activities. What is special among them is the chain-linked model that has a greater significance in guiding enterprises to conduct technological innovation activities. According to this model, an enterprise—when conducting technological innovation activities—not only needs to achieve better internal integration among various functions but also needs to broaden its connections with the outside in order to obtain external resources; only in this way can an enterprise better improve the efficiency and rate of success in its technological innovation activities.

Many scholars in China followed the abovementioned technological innovation theories and conducted studies in multiple aspects about how to enhance the technological innovation capabilities of the Chinese enterprises. Chen (1994) proposed the learning model explicating the process for Chinese enterprises to go from technology import to independent innovation. Gao and Fu (1996) conducted a survey of 1051 enterprises and then analyzed the seven key issues of technological innovation in Chinese enterprises. In recent years, more scholars have demonstrated the importance of technological innovation in building

[1] In that model, the two scholars proposed that there are five paths to technological innovation: "[the first] path of innovation starts from the phase of invention as well as analysis and design, then goes through detailed design, test, production and then to distribution and sales—this is the chain at the center of technological innovation; the second path is a series of feedback going from market to the center chain, among which the one called potential-market-to-center-chain feedback is of special importance; the third is the chain linking research with knowledge and with the center chain; the fourth is the path going from research for new knowledge generation to fundamental innovation; the fifth is the path going from innovative product to scientific feedback and to improvement of innovation." This model not only emphasizes the role of the push from technology but also, more importantly, highlights market as the dominant factor in the innovation process and makes clear that the primary innovation path starts with the market and ends with the market.

enterprise competitiveness (Zhao et al., 2001; Liu et al., 1997).[1] Chen (2001) pointed out that technological innovation has become an important way for technological progress and economic development and that for an enterprise, the ability to carry out technological innovation activities will be the key factor in gaining competitive advantages. Huang (1997) demonstrated that market innovation is the key factor in deciding the fate of a company and discussed how different enterprises should make strategic choices of market innovation domain and market innovation point. Wu (2000) proposed three strategic models of technological innovation in Chinese enterprises[2], that is, the introduction, digestion, and absorption innovation model; imitation strategy model; and market segmentation, cost minimization, and technological advancement model. From the perspective of technological innovation strategy, Wei (2001) then pointed out that the internal qualities based on which an enterprise gains competitive advantages are its technological capabilities and core competency, described the typical ways by which such capabilities grow with their growth trajectories, and put forward measures by which Chinese enterprises can enhance their technological innovation capabilities. From the abovementioned studies conducted by the domestic scholars, emphasis has been placed on strategic studies of technological innovation to a great degree. Then, how to formulate technological innovation strategies? It is necessary for us to first review the evolution of related strategic ideas.

Chandler (1962) and Ansoff (1965) proposed the strategic idea with an analytical framework that is resource-based and emphasizes the comparison between internal strengths and internal weaknesses of

[1] Zhao, Cai, and Luo (2001) proposed the model for building internal core competitiveness of an enterprise through self-development and the model for building core competitiveness of an enterprise through M&As and through forming strategic alliances with enterprises with complementary strengths.

[2] The technological innovation strategy of an enterprise refers to an overall design or a design affecting the overall situation of the technological innovation activities in an enterprise (Wu, 2000).

an enterprise. In the analytical framework, they particularly stressed the great significance of best behaviors, internal resources, and internal competitiveness of an enterprise for its success. In the 1970s, Western management scholars introduced the analytical method of the industrial organization theories into the field of strategic management, and proposed the structure-conduct-performance framework of strategic analysis; Porter (1980, 1985), in his analytical framework, pointed out the important role of technology in competition and linked technology to the five driving forces that promote industry-wide competition. Porter's analytical framework places the technological activities of an enterprise in the broader setting of industry-wide competition and his theory was once a basic analytical framework in the field of strategic management in the 1980s. However, the thinking proposed by Chandler (1962) and Ansoff (1965) in strategic management is—to a large extent—a theory of rational strategic thinking, which underestimates the role of technological changes and market competition. To a large extent, with the ever-increasing globalization of world economy, the external environment has become increasingly complex and the change in it is ever more dramatic, and as a result, it is impossible for managers to have an accurate understanding of this complex environment and to make predictions about its change; thus, the singular focus on concentrating massive resources to create monopoly power will tend to ignore lucrative emerging markets and the ever-growing needs of customers (Xu, 2000). Porter's (1980, 1985) analytical framework also has the deficiency of underestimating the role of technology change and market competition, and the framework, when conducting strategic analysis, has clearly limited its attention to industry-wide development.

With dramatic changes in the international economic environment, as well as early signs of a knowledge-based economy, Mintzberg and Quinn (1992) proposed the idea of gradual progression as the strategic management thinking. The core of this strategic thinking is that humans with limited rationality cannot fully understand the complex and ever-changing environment, and thus in strategic management, we should follow the idea of gradual progression and should not rely on rationality and that it is necessary to continuously revise—with new circumstances, new information, and new understandings—the environment, strengths, weaknesses, and directions for change in an enterprise and to adjust the existing strategies accordingly. However, Mintzberg and Quinn did not give details about how an enterprise can continually adjust its strategy and in what way an enterprise can adjust and realize its strategic advantages.

Barney (1991) and Conner (1991) et al. again put forward the idea that RBV[1] should be emphasized in strategic management. Chen (1996) and Gimeno and Woo (1996) put forward a strategic thinking based on the enterprise technological capabilities with an emphasis on dynamic competition, the enterprise itself, and boundary with the business environment. Prahalad and Hamel (1992) suggested that the essence for managing an enterprise from the strategic perspective is to place the emphasis on the technological capabilities and reserves of the enterprise; the key factor for an enterprise to gain technological advantages, especially strategic advantages, and to achieve sustained development autonomously is to have the ability to effectively use and control technological knowledge. They proposed that the core capabilities of an enterprise exist in the product-technology platform—the core product—and that only after building such a platform can rapid technological innovation be achieved. Meyer and Utterback (1993) argued that the core competency of an enterprise should refer to the combination of its R&D capabilities, manufacturing capabilities, and marketing capabilities and that it is an important factor affecting business performance. Leonard-Barton (1995) argued that core competency is the knowledge system, which is unique to the enterprise and brings competitive advantages to the enterprise, and that in order for an enterprise to deliver continued autonomous innovation, there must

[1] RBV: resource-based view.

have been continued accumulation of core competencies. Conner (1991) argued that the resources constituting the core competency must be the resources that are scarce, unique, inimitable, and non-transactional; only when an enterprise is in possession of such resources can it claim a unique position in its industry. Coombs (1996) argued that the core competency of an enterprise includes technological capabilities and organizational skills that can combine such capabilities into effective competencies. In addition, Coombs emphasized that as the carrier of core competency, talent is the most important resource of an enterprise.

For an enterprise's technological innovation activities, the change in strategic thinking from emphasizing enterprise technological development and market push to building upon core competency is very important for technological innovation management. Due to the complexity and unpredictability of environment changes, theories based on the environment analysis and driving forces have run into unprecedented difficulty and traditional ideas in strategic management have come to a difficult and critical stage; meanwhile, a group of enterprises have relied on strategic management that is based on accumulating their own capabilities and have been able to gain competitive advantages in a complex and ever-changing environment. In this era of knowledge economy, the environment is undergoing dramatic changes and strategic management that is based on core competency will become the foundation, upon which an enterprise can gain advantages. The theory of core competency has changed the adaptation view, which has been in existence as the strategy theory for a long time, and it has, by strategic extension, changed the strategic posture from a defensive one to an offensive one. By extending the core competency, an enterprise can change its environment in the industry and its own posture of development. This theory demonstrates that with regard to environment changes, an enterprise is not passive and powerless; strategic management thinking based on core competency shifts the emphasis from the mix of operations to the mix of capabilities and from market share to opportunity share (Xu, 2000).

At present, the trend of global economic integration is moving faster and faster and internationalized production and trade is growing at an unprecedented speed. Along with such developments, the internationalization of technological innovation and R&D is getting more and more attention from governments and enterprises.

Especially in the last decades, due to worldwide rapid development, multinationals have become the promoter for internationalization of technological innovation and R&D (Patel and Pavitt, 1991; Cantwell, 1995; Nonaka & Takeuchi, 1995; Robert, 1995a, 1995b); consequently, the internationalization of technological innovation and R&D is becoming an ever more important factor in raising the core competitiveness of an enterprise and strengthening its global competitive position.

Multinationals are the product of internationalized production and advanced technological development in the world. Ever since the 1990s, with the in-depth development of economic globalization, multinationals have been expanding effectively around the globe in an unprecedented speed and scale; with their unique international networks, their technological innovation activities have been developing rapidly worldwide; and relying on the world market and global talent, they have reaped a great deal of economic and social benefits. It can be said that it is on the foundation of the development of multinationals that internationalization of technological innovation can develop continuously and push the studies on the management of technological innovation to a new phase.

Internationalization of technological innovation is characterized by globalization of innovative sources, diversification of innovative talent, and networking of the various organizations of technological innovation; it has changed the invariable situation where an enterprise always obtains the necessary technological knowledge internally and has brought about the formation of an enormous transaction network linking the internal system of the enterprise with the various external organizations. An enterprise can draw from a broader range of external knowledge and data sources to build up its own technology and knowledge structure. Innovation management and knowledge integration have become the key to success, and technology alliance and innovation outsourcing have become preferred choices for enhancing the capabilities for technological innovation and for responding to competitive pressure brought about by a knowledge-driven economy. It can be predicted that internationalization of technological innovation for sure will become the top priority in the enterprise development of various countries and that the scale, trend, and impact of such internationalization will be worldwide.

China has already joined the WTO and, now as a part of the global economy, will be strongly influenced by this trend of internationalization of technological innovation. At this stage, most Chinese enterprises are limited by their own individual strengths and the national conditions, thus are basically in a blank or early stage with respect to internationalization of technological innovation, have not acquired in-depth understanding about the various aspects of internationalization of technological innovation, and lack the necessary experience to execute internationalization of technological innovation. In this day of ever-intensifying international economic competition, being faced with internationalized R&D, production, and trade strategies implemented by the multinationals of different countries, in order to develop a foothold in competition, Chinese enterprises must adapt to the demands of a globalized economy and internationalized competition as soon as possible, formulate their own strategy for internationalization of technological innovation, break through the original strategic thinking in the management of technological innovation that is based on the industry and environmental analyses, and achieve the transition toward the strategic thinking by strengthening the core competencies of the enterprise, so as to be able to participate in the international competition on a more solid foundation.

3.2 Analysis of the state of Chinese enterprises on internationalization of technological innovation

In early 2002, we issued a total of 279 questionnaires on internationalization of technological innovation and R&D activities to China's well-known large enterprises and high-tech enterprises. At the end of August 2002, we received a total of 28 questionnaires. The recovery rate (i.e., about 10.04%) was quite low. One main reason for the low recovery rate of questionnaires is the low degree of usage of internationalization of technological innovation in Chinese enterprises—many enterprises did not carry out activities of internationalization of technological innovation and thus they simply could not fill out our questionnaires. Other than that, some enterprises were in industries that were fiercely competitive; thus, for reasons of confidentiality, some enterprises were not willing to fill out our questionnaires. Out of those that returned the questionnaires, we selected nine enterprises and, through interviews with the senior leaders and technology directors, gained further understanding of the state of their internationalization of technological innovation.

3.2.1 Scale and distribution of R&D funding and types of internationalized innovation activities

The scale of R&D funds and the establishment time of the R&D organizations for the 28 enterprises surveyed as of August 2002 are as follows. The earliest of the surveyed enterprises first established overseas R&D branches between 1985 and 1989, but such enterprises account for only 10% of those surveyed; most established overseas R&D branches between 1995 and 1999, and they account for 60% of those surveyed; and the remaining 30% established overseas R&D branches after 2000. This shows that internationalization of technological innovation by Chinese enterprises is still at an early stage (Figure 3-1).

Among the surveyed enterprises, in 2001, 33% spent less than 5 million yuan (RMB), 44% spent between 5 and 20 million yuan, 22% spent between 20 and 50 million yuan, and none spent more than 50 million yuan on overseas R&D branches. Surveyed enterprises mainly spent between 5 and 20 million yuan on their overseas R&D branches (Figure 3-2). The scale of R&D expenditures by Chinese enterprises was very low compared to that of multinationals. When asked about future overseas R&D expenditures, most of the enterprises indicated that they would increase the investment for overseas technological innovation and R&D activities. According to the survey data, by 2005, 66% of enterprises estimated that their overseas R&D investment would exceed 20 million yuan and half among the 66% would invest more than 50 million yuan for overseas R&D branches (Figure 3-2). Forty percent of the overseas R&D branches established by Chinese enterprises are based in the United States, 13.3% in Europe, and 0.7% in Japan, which is consistent with the findings of Gerybadze and Reger

(1999) and Gassmann and Von Zedtwitz (1999).[1] Twenty-seven percent of the overseas R&D institutions established by Chinese enterprises are based in other areas; this is mainly due to the fact that the international sales and production of some of the Chinese enterprises have gradually extended to some third world countries and, as a result of such extension, these enterprises need to provide technical services and technical support for production and sales in these regions. For the next five years, the surveyed enterprises planned to slightly increase the proportion of R&D branches in Europe, Southeast Asia, and other regions and, at the same time, reduce the proportion in the United States (Figure 3-3).

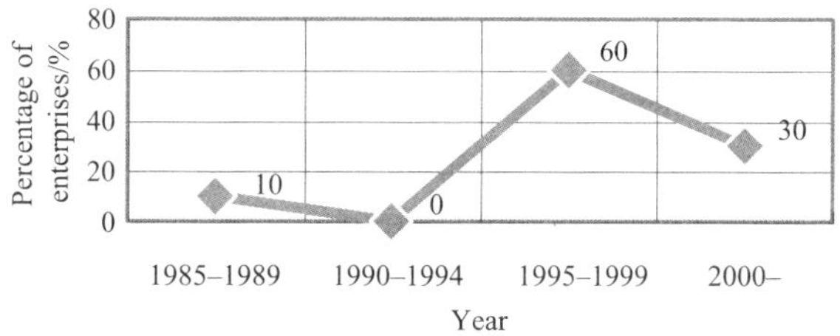

Figure 3-1 Distribution by establishment time of overseas R&D branches

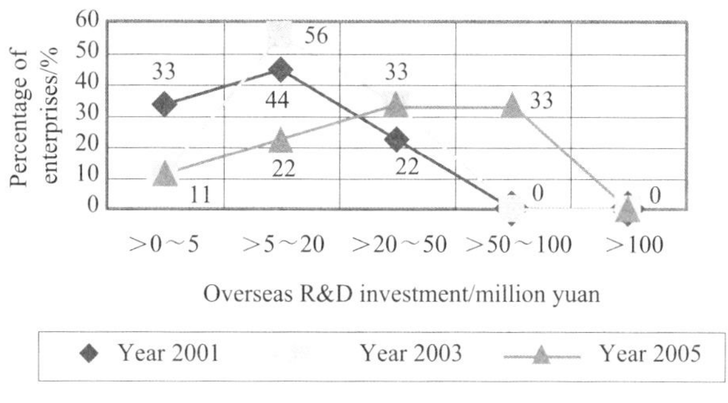

Figure 3-2 Funding for overseas R&D

[1] Studies by Gerybadze and Reger (1999) and Gassmann and Von Zedtwitz (1999) indicated that investments by multinationals also mainly concentrate on the triangular region, that is, the United States, Europe, and Japan.

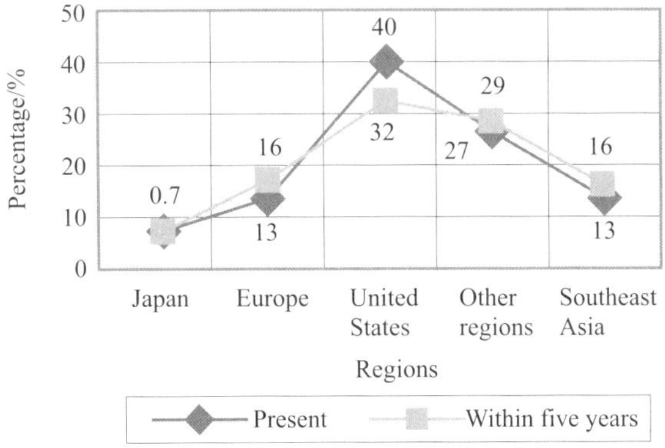

Figure 3-3 Percentage of overseas R&D branch locations

In our survey, we classified internationalized technological innovation activities into five types: A—basic scientific research (long term); B—applied research based on basic research, creating possible commercial applications (medium term); C—market-oriented commercial product development (short term); D—improvement of product development (short term); and E—market-oriented reformation of technological processes (short term). Our survey shows that the strategic intent of Chinese enterprises to develop internationally is—through overseas expansions and establishment of overseas R&D institutions—to develop potential technological capabilities, improve core technological competencies, and eventually achieve the objective of developing and stabilizing domestic markets, while taking part in the overseas market. Their overseas technological innovation activities are mainly aimed at market-oriented commercial development and improvement of product development as well as market-oriented reformation of technological processes. The improvement of product (short term) is placed after that, and basic scientific research is given the least important position. It is predicted that, over the next five years, as overseas activities further expand, product development, applied research, and basic research will gain more importance in the overseas R&D activities of an enterprise; however, enterprises surveyed are still extremely focused on market-oriented commercial product development and short-term product development, where basic research is given the least priority (Figure 3-4).

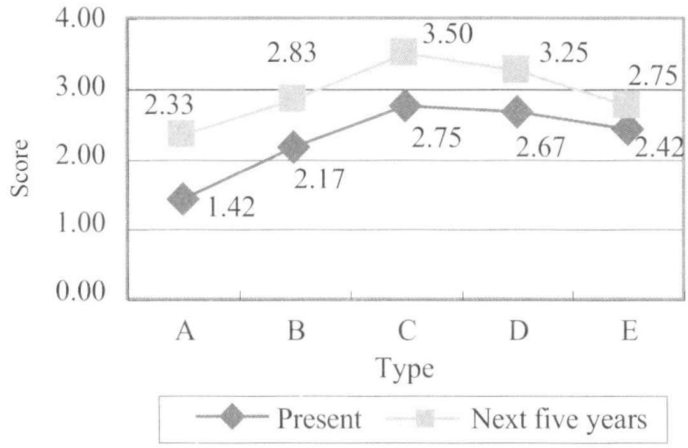

Figure 3-4 Types of activities conducted by overseas R&D branches

3.2.2 Strategic factors for implementing overseas technological innovation

The factors for technological innovation and its internationalization are quite complex. Of course, the form and degree of internationalization of technological innovation change over time; accordingly, the research conclusions will be different, and, for different countries and different enterprises, the driving factors for internationalization of technological innovation are also different. In the survey, we identified 10 driving factors for the internationalization of technological innovation, from A to J: A—responding to competition and implementing enterprise development strategy; B—expanding and enhancing the enterprise potential technology capabilities; C—providing local subsidiaries with necessary technology support in product, technological process, and so on; D—pushing new products into local markets as soon as possible; E—making use of good local R&D hardware and infrastructure; F—taking advantage of local preferential policies; G—setting up technology monitoring points, getting to know about competitors, and determining the need for technology development; H—recruiting outstanding R&D talent and absorbing advanced technologies and scientific knowledge from abroad; I—cooperating with the local enterprises or research institutes to work on R&D projects; and J—getting close to suppliers. Through our investigation, we find that, the driving factors for internationalization of technological innovation of Chinese enterprises are mainly as follows: responding to competition and implementing enterprise development strategy, expanding and enhancing potential technological capabilities, recruiting outstanding researchers, and absorbing advanced scientific and technological knowledge from abroad. We find that the market factor—which is deemed by many scholars as very important in the process of internationalization—is not an important factor for Chinese enterprises (Figure 3-5).[1] We analyzed and concluded that this is mainly due to the fact that the level of internationalization by Chinese enterprises is still relatively low and that sales and production activities of some enterprises have not yet reached the international market. In addition, Chinese enterprises are mainly focused on domestic market, which is large enough for them. Under such circumstances, the enthusiasm for these enterprises to expand into the market overseas is not very high, which is consistent with the views of Gerybadze and Reger (1999). It should be noted that, for the surveyed enterprises, taking advantage of local preferential policies and getting close to suppliers are the least important of all driving factors.

[1] The market in our questionnaire mainly refers to the international market outside of the home country.

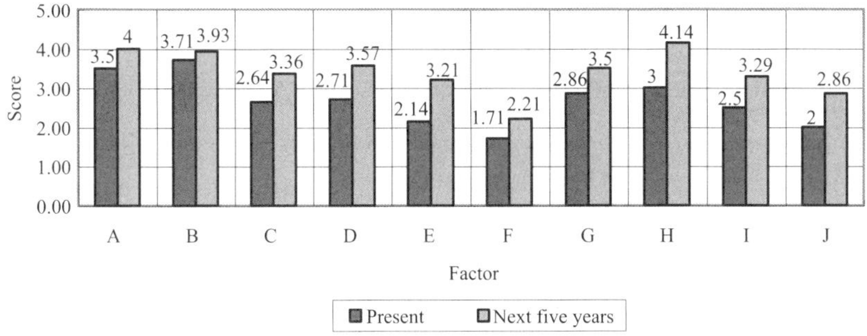

Figure 3-5 Driving factors for internationalization of technological innovation

Chinese enterprises are also aware of their challenges and competition, which come along with China's accession to the WTO, and they have recognized the importance of the international market for enterprise development. Thus, according to the responses from the surveyed enterprises when asked about the driving factors for their efforts to internationalize technological innovation in the next five years, responding to competition and implementing enterprise development strategy, expanding and enhancing the enterprises' potential technological capabilities, and recruiting outstanding R&D talent and absorbing advanced technologies and scientific knowledge from abroad are still ranked first, whereas pushing new products into local markets as soon as possible and providing local subsidiaries with necessary technology support in products, technology processes, and so on have become important factors immediately after those ranked first.

3.3 Strategic framework of Chinese enterprises for internationalization of technological innovation

The strategic factors in internationalization of technological innovation are quite complex, and there is a considerable amount of literature on this subject. Mansfield et al. (1979) once conducted a study on overseas R&D activities by U.S. companies, and the results showed that "the main purpose for a majority of the companies is to carry out technology transfer." Ronstadt (1977, 1984) came to a similar conclusion: U.S. multinationals set up R&D units abroad to carry out technology transfer and to adapt their technologies to the needs of the local markets. Hewitt (1980) found that motivational factors for overseas R&D activities can be classified into three categories: those for adaptation, those for location-based market, and those for globalization. Similarly, Patel and Vega (1999) summarized the motivational factors for multinationals to internationalize technological innovation into three types: making adjustments to products, technological processes, and raw materials in order to adapt to the foreign market, and providing technical support for overseas manufacturing units; tracking dynamic changes in overseas technological development; and developing core products and key technologies abroad.

With the development of internationalization of technological innovation activities, the motivational factors behind such activities are becoming ever more complex. Niosi and Godin (1999), by their research, showed that commercial and pre-commercial R&D activities as well as product development activities are the most important driving factors behind setting up overseas subsidiaries. Other noteworthy factors include supporting the parent company's marketing efforts in the host country, providing the parent company with technology support, assisting the parent company in transferring technology, and so on. The secondary tasks of overseas research institutes are to provide technology support for external companies or organizations, to provide technology support for the government, to carry out basic research, and to conduct other studies.

By investigating and studying foreign R&D institutions in the United States and the R&D activities abroad by U.S. multinationals, Dalton and Serapio (1999) systematically summarized the motivational factors driving overseas R&D activities by U.S. multinationals and R&D activities in the United States by foreign multinationals toward the end of the 20th century. They divided the motivational factors driving overseas R&D activities by U.S. multinationals into two classes: the market- or demand-driven class and technology- or supply-oriented class. They further divided these factors into the following: developing and designing products suitable for the local market; developing new products for the local market; supporting the production, sales, and service units of the parent company in the host country; mining R&D talent; developing new science and new technology; developing new products for global customers; monitoring technological development overseas; and participating in joint or cooperative research.

Gerybadze and Reger (1999) proposed a new model of internationalization of innovation. On the basis of the combination of market and technology, they suggested that there are four models of internationalization of innovation, that is, innovation based on science and research when there is a large R&D base in the home country (Type A), innovation based on science and research when there is a small R&D base in the home country (Type B), R&D closely coupled with the leading and major market in the home country (Type C), and R&D closely coupled with the leading overseas market (Type D). The authors concluded that major countries themselves often hold on to important markets, but small countries often hold on to small markets and weak R&D bases. Then, they further pointed out that some companies are driven by innovation based on scientific research and such companies tend to set their R&D laboratories in close proximity to places with the best intellectual resources, regardless of whether such places are located at home country or abroad. The innovation-related characteristic of Type A companies is a strong research base in the home country. Such companies are not very motivated to go to other regions of the world to seek the fruits of science and the research talent in their fields. However, Type B companies are particularly motivated to get close to S&T centers in other countries, in an effort to obtain advanced S&T. Type C and Type D companies particularly emphasize downstream links and the close combination of the leading market, R&D, and product innovation; they place their emphasis on following customer needs and innovation positioning; some companies seek those lead customers in their home country, but they are likely to be subject to the size of the domestic market. Companies from major countries enjoy both developed domestic market and strong research base in their home country, and thus, they will seek suitable innovative environment and will place special emphasis on integrating their activities in the host country with those of the company's headquarters; consequently, overseas innovation activities are often supplementary. Multinationals from small countries and those from large countries without a developed domestic market, such as European companies, place their main force on R&D and innovation activities in the United States and Japan—two developed markets. Type D innovation activities will continue to gather investments overseas and develop core competencies.

Von Zedtwitz and Gassmann (2002) put forward a model for the evolution of an enterprise's R&D activities. With this model, the authors suggested that a company's R&D activities would develop two dimensions along research or development: the first is science-driven and the second is market-driven. One possible evolutionary path for an enterprise's R&D activity is to develop in the home country before turning into technology-driven internationalized R&D activity and eventually evolving into globalized R&D activity. The other possible evolutionary path is to develop from the home country model to market-driven internationalized R&D activity and eventually also to globalized R&D activity. The point of the first R&D evolutionary path is first to expand the market and then to develop cutting-edge S&T, and that of the latter is first to develop cutting-edge S&T and then expand into the international market. It is certainly not true that the R&D activities of every enterprise can develop into the globalized phase; the development of R&D activities is restrained by the industry and the organization's resources.

After summarizing the aforementioned studies by the foreign scholars, we come to believe that the factors driving internationalization of technological innovation are complex and, to different countries and to different enterprises, the driving factors for the technological innovation strategy are different. Therefore, we need to consider the actual situation of China and formulate and adjust the strategy for internationalization of technological innovation with the focus on the actual situations in the Chinese enterprises. We believe that the model for internationalization of technological innovation formulated by Von Zedtwitz and Gassmann (2002) and Gerybadze and Reger (1999) has a guiding significance for us in developing strategic framework for internationalization of technological innovation by Chinese enterprises[1] and that, equally, the research conclusion that we described in the last section on core competency also has important guiding significance for us in developing the strategic framework for internationalization of technological innovation by Chinese enterprises.

The strategy for internationalization of technological innovation refers to an overall design or a design affecting the overall situation of internationalization of technological innovation activities conducted by an enterprise. This strategy is the path toward internationalization of technological innovation activities by an enterprise, the blueprint for the enterprise's internationalization of technological innovation, and the enterprise's strategic plan for carrying out technological innovation activities. An enterprise should rely on such strategies to establish its loyalty to the customers and to gain a sustainable competitive advantage over its competitors. Strategies for internationalization of technological innovation must integrate the decisions by various departments, management personnel, and technical personnel and the actions taken by them into a coordinated strategic plan.

The strategic objective of the strategic framework that we proposed for internationalization of technological innovation based on core competency requires that Chinese enterprises achieve globalization in the not-so-distant future in both the technology and market dimensions (Figure 3-6). In other words, it is hoped that each enterprise's technological innovation activities will be extended to every corner of the world, so as to diversify the sources of technological knowledge and obtain the knowledge and technology required by the enterprise's technological innovation from the various knowledge and technology centers of excellence worldwide and that each enterprise's products will be pushed to the global market so as to ultimately achieve the coordinated development of enterprise core capabilities and enterprise global business operations.

[1] However, the model proposed by Gerybadze and Reger (1999) is mainly aimed at the dynamic and short-period innovation.

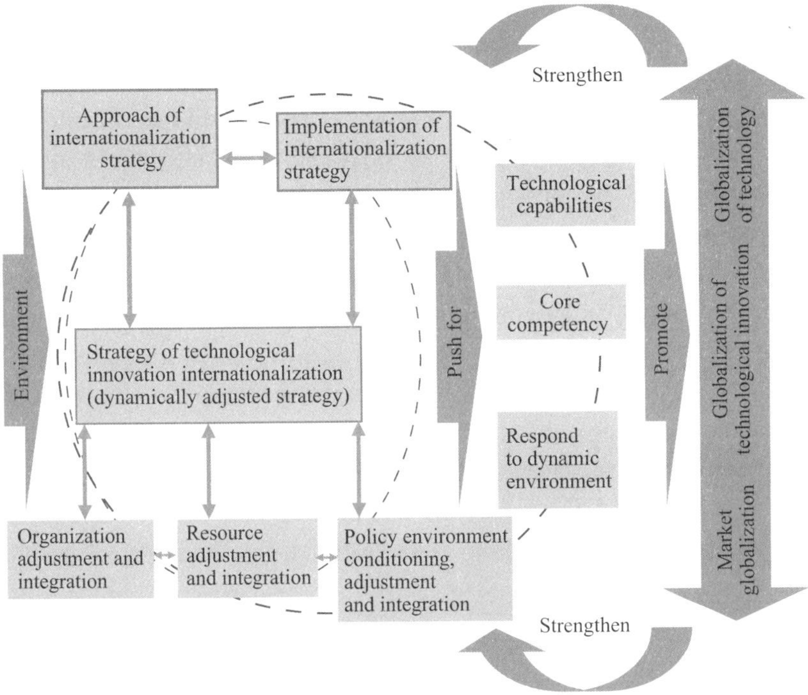

Figure 3-6 A strategic framework for internationalization of technological innovation by Chinese enterprises

The guiding idea of our proposed strategy is the guiding strategic thinking based on the constructive core competency of an enterprise, which requires that enterprises rely on the accumulation of their own capabilities—by improving the enterprise's technological capabilities and core competencies—respond to the complex and changing environment, and achieve sustainable competitive advantages in relation to competitors. In this day of knowledge economy, the environment is undergoing dramatic changes and, thus, strategic management based on core competency will become the foundation, upon which an enterprise can gain advantages. The theory of core competency has changed the adaptation view, which has been in existence as the strategy theory for a long time, and it has, by strategic extension, changed the strategic posture from a defensive one to an offensive one.

Our strategy for internationalization of technological innovation based on core competency attaches great importance to the constructive core competency of an enterprise itself and to the integration of the internal resources, organizations, and policies. Thus, our strategy is to integrate the decisions by various departments, management personnel, and technical personnel with the actions taken by them into a coordinated strategic plan. Our strategy must closely match the integrated organizational strategy, resource strategy, and environmental strategy of the enterprise. When an enterprise's technological innovation becomes internationalized, even if it is still only at the earliest stage of internationalization, profound changes will happen following the enterprise's internationalization management of technological innovation; thus, without proper adjustment and integration of an enterprise's organizations, resources, and policies as well as strategies, it will be impossible to achieve the strategic objective of internationalization of technological innovation. In this sense, the adjustment and integration of an enterprise's organizations, resources, and environment policies constitute a key element for implementing the enterprise's strategies.

An enterprise's technological innovation strategy based on its core competency emphasizes that the enterprise should build its core competency so as to be able to respond to the complex and changeable environment. However, this does not mean that we can ignore business environment analysis. We still have the belief that the influence of external changes to an enterprise in its industry and business environment is important. Thus, we still have to perform a detailed analysis of its external environment in order to cautiously adjust our strategy for internationalization of technological innovation, and only by doing so can we ensure that the enterprise achieves its strategic objective of internationalization of technological innovation. When implementing the technological innovation strategy, we must carefully analyze the resources and strengths of the competitors and the strengths and advantages as well as disadvantages of the enterprise's resources and choose the appropriate strategic path to gradually achieve its strategic objectives.

With such a strategy, our goal is to achieve globalization in both dimensions—technology and market. The central aim is to improve core competencies of Chinese enterprises, and thus, while implementing the strategy, we must keep the emphasis on expanding and enhancing the enterprises' potential technological capabilities. With this strategy, the enterprise must provide local subsidiaries with necessary support in products, technological processes, and so on; push new products into the local market as soon as possible; make good use of local R&D hardware and infrastructure; take advantage of local preferential policies; learn about competitors and determine the need for technological development; and recruit outstanding R&D talent and absorb advanced technologies and scientific knowledge from overseas. Meanwhile, in the process of technological innovation, an enterprise must attain the following one by one along its strategic development path: market-oriented commercial product development (short term); improvement on product development (short term), and market-oriented reformation of technological processes (short term); applied research that is based on basic research, creating possible commercial applications (medium term); and basic scientific research (long term). It is worth noting that in our strategic framework for internationalization of technological innovation, the building and strengthening of core competency and the degree of internationalization of technological innovation are complementary to each other. Here is the interaction between them: on the basis of existing core competency, internationalization of technological innovation moves forward; in turn, internationalization of technological innovation will strengthen the core competency; and once core competency is strengthened, it will further promote internationalization of technological innovation.

3.4 Strategic approaches of internationalization of technological innovation for Chinese enterprises

With the survey of domestic enterprises and consideration for the development process of foreign multinationals and the successful practice in the internationalization process of some domestic enterprises, we suggest four strategic approaches to the Chinese enterprises for internationalized technological innovation activities. These four strategic approaches evolve from the lowest level to the highest level, and the focus is on the ongoing evolutionary process from domestic innovation to international innovation (Figure 3-7). These four strategic approaches are derived from the strategic framework that we proposed earlier for internationalization of technological innovation. Analyzing the scale, strengths, and business environment and adopting a suitable strategic approach on the basis of such analyses will help Chinese enterprises carry out internationalization of technological innovation successfully and improve core competencies as fast as possible.

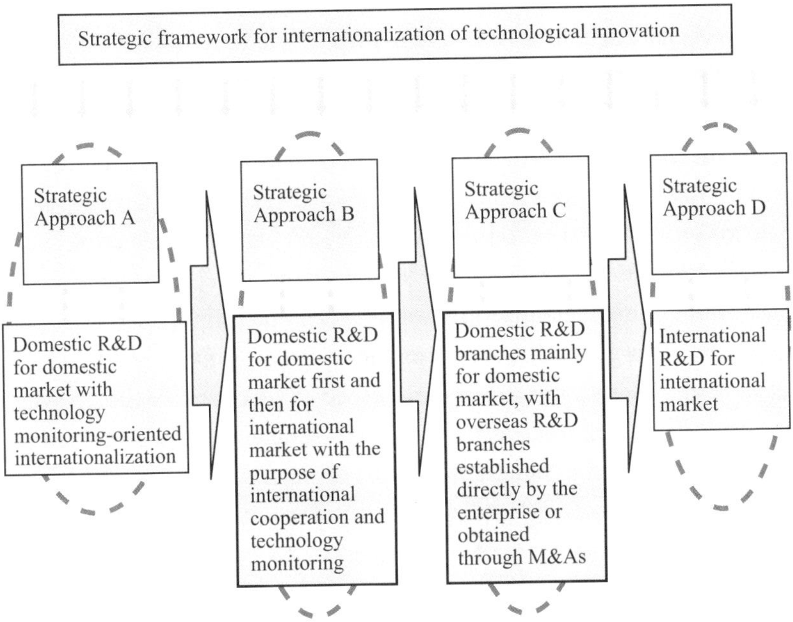

Figure 3-7 Evolution of strategic approaches for internationalization of technological innovation

3.4.1 Strategic Approach A: domestic R&D for domestic market with technology monitoring-oriented internationalization

One way to enhance an enterprise's technological innovation is using a technology monitoring strategic approach for the domestic market with domestic R&D—Strategic Approach A (Figure 3-8). This strategic approach is suitable for enterprises with limited scale and limited strength, at a low degree of internationalization in production and sales, with a very limited number of products sold to foreign markets. For such enterprises, domestic markets are more important than foreign markets. Foreign enterprises, such as ABB and GE, adopted similar strategic approaches in the early stages of their development.

With this strategic approach, the strategic direction of an enterprise is very clear: it is to carry out technological innovation activities for the domestic market and to seize the domestic market. However, to meet the needs of consumers in the domestic market, especially after China's accession to the WTO, and to face the competition in the domestic market from foreign manufacturers, the technology and knowledge that can be acquired domestically are insufficient. Just by relying on technological innovation that is based on the domestically available technology and knowledge, an enterprise cannot respond adequately to competition from foreign manufacturers. Therefore, such enterprises need to set up places overseas for technology monitoring, so as to know the trends of technological developments overseas, especially the trends of technology and knowledge and product technological processes that are adopted by foreign enterprises in competition with Chinese enterprises in China's domestic market. By establishing technological monitoring points overseas, providing feedback to domestic R&D units, and planning R&D projects in the domestic R&D units, an enterprise can improve its potential technological innovation capabilities and enhance its ability to compete in the domestic market.

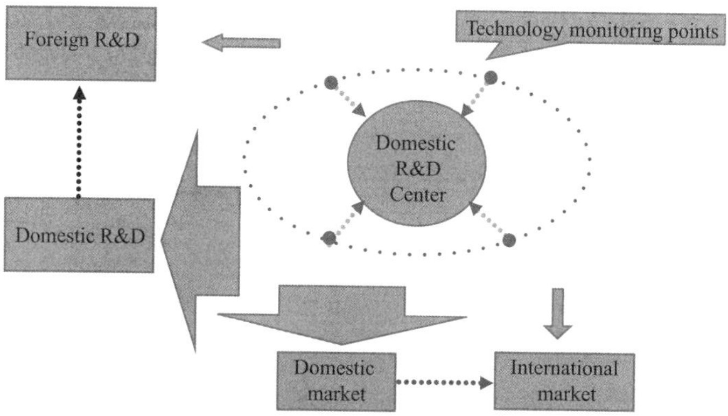

Figure 3-8 Strategic Approach A of internationalizing technological innovation for Chinese enterprises

With this strategic approach, the immediate goals for internationalization of enterprise technological innovation are to set up multiple technology monitoring points overseas and to make arrangements to ensure as much as possible that there is effective communication among such technology monitoring points to improve the efficiency of technology monitoring and information feedback. A medium-term goal is—on the basis of domestic R&D—to strive for cooperative project development inside and outside of the home country with foreign enterprises and foreign R&D institutions, so as to further improve the enterprises' technological innovation capabilities and, at the same time, look for the right opportunity to push their products to overseas markets. It is worth noting that enterprises following this strategic development approach should attach great importance to technological innovation capabilities in developing market-oriented commercial products (short term), improving product development (short term), and reforming market-oriented technological processes (short term). Though implementing the strategy for internationalization of technological innovation is good, the enterprises should not place too much pressure on their monitoring and R&D capabilities to keep track of the medium- and long-term basic S&T.

3.4.2 Strategic Approach B: domestic R&D for domestic market first and then for international market with the purpose of international cooperation and technology monitoring

The Chinese enterprises following Strategic Approach B (Figure 3-9) should be greater in terms of size and strength than those following Strategic Approach A. Generally, the enterprises of this strategic approach have already achieved a certain level of internationalization in sales and have already started (or are preparing to start) building their own production facility or subsidiary overseas. For such enterprises, the domestic market is still the most important one; however, medium- and long-term developments demand that such enterprises strive to expand into foreign markets because such foreign markets constitute an important factor for the further development of such enterprises.

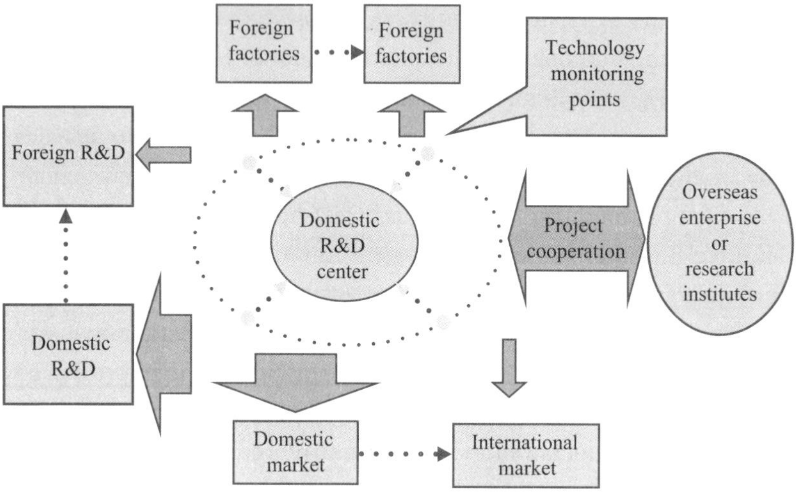

Figure 3-9 Strategic Approach B of internationalization of technological innovation for Chinese enterprises

With Strategic Approach B, the short-term goal for the enterprises is to make progress in the internationalization of both technology and market, but the strategic focus stays on domestic R&D and market. The strategic priorities are as follows: developing and strengthening the enterprise's potential technological capabilities, setting up technology monitoring points, learning about the competitors, and determining the trend of technological development; joining the local enterprises and research institutes in cooperative R&D projects and absorbing advanced technology and scientific knowledge from abroad; and strengthening and expanding in the domestic market and, at the same time, striving to provide local subsidiaries and production departments in the overseas market with necessary technology support in products, technological processes, and so on. The medium-term goal under this strategy is to further expand the scope of technology monitoring, extend the range of technology cooperation, and strive to establish R&D centers abroad, in order to increase the strength and depth of an enterprise's internationalization of technological innovation.

The practices of such internationalization of technological innovation are to establish overseas technology monitoring points. While monitoring the latest knowledge and technology abroad, the enterprise should pay close attention to the needs of overseas consumers, and send as much as possible the information about knowledge, technology, and consumers to the enterprise's headquarters. An enterprise can carry out cooperative development with overseas enterprises through joint projects; such cooperative development may be done at home or overseas. An enterprise can enhance its own technological innovation capabilities as much as possible by means of developing cooperative projects. For an enterprise following this strategic approach, in the process of technological innovation, its domestic R&D headquarters plays the central role. Whether it is the technology and knowledge feedback from technology monitoring points or project cooperation with overseas enterprises or R&D institutions, such activity is conducted with the support and under the control of the R&D center in the home country. Providing technology support to overseas subsidiaries and production departments and providing mature product prototypes to overseas manufacturing departments are also done by the R&D center in the home country.

3.4.3 Strategic Approach C: domestic R&D branches mainly for domestic market, with overseas R&D branches established directly by the enterprise or obtained through M&As

Enterprises adopting Strategic Approach C (Figure 3-10) are the ones with a considerable amount of strength and size. An enterprise following this strategic approach can not only set up overseas technology monitoring points and conduct cooperative research with foreign R&D institutions but also perform more actively to directly establish R&D branches and to acquire R&D institutions by M&As. An enterprise following this approach can obtain more timely and effective knowledge and technology than those following the previous two strategies, integrate such knowledge into its core competency more effectively, and improve its technological innovation capabilities, so as to achieve the strategic goal of securing a firm hold on the domestic market and to take part in the international market.

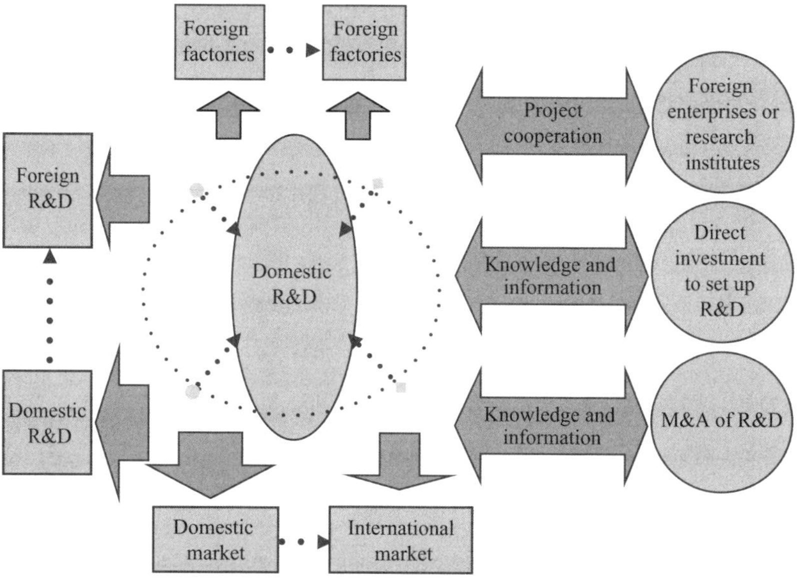

Figure 3-10 Strategic Approach C of internationalization of technological innovation for Chinese enterprises

Its strategic priorities include the following: to develop and enhance the enterprise's potential technological capabilities; to set up technology monitoring points and overseas R&D branches, understand competitors, and determine the need of technology development; to absorb advanced technology and scientific knowledge from overseas R&D branches; and to more effectively improve its technological innovation capabilities through cooperative R&D projects with enterprises or research institutes overseas. Thus, it is suggested that Chinese enterprises provide necessary technology support for local subsidiaries in products and processes, introduce new products into the local market as soon as possible, and recruit competent R&D researchers and management staff overseas to attain internationalization of technological innovation.

For an enterprise following this strategic approach, the domestic R&D center still takes a central position in its internationalization process of technological innovation. Such an enterprise guides the R&D activities at the overseas branches and establishes cooperative relationships with overseas R&D institutions through its domestic R&D center, and the overseas technology monitoring points constantly provide feedback about knowledge, technology, and local consumer needs to the domestic R&D center, while some overseas sales networks also provide feedback to the domestic R&D center. Such enterprises attach great importance to establishing R&D branches overseas by direct investment and to acquiring—by the operation of overseas capital market—medium- or small-sized companies, which are necessary for extending the technological capabilities of the enterprises.

The medium- to long-term objectives for the Chinese enterprises following this strategic approach include the following: to obtain more R&D branches overseas, especially knowledge centers, technology centers, and other centers of excellence, by direct investment and M&As; to integrate overseas R&D branches into a more efficient network through organizational integration, resource integration, and policy integration; and to weaken the central position of the domestic R&D center. Therefore, overseas R&D centers will become more autonomous and independent, the enthusiasm of overseas branches will be fully mobilized, and the creativity of R&D staff in these branches will be fully motivated. To achieve these objectives, the integration of the network is crucial.

Enterprises following this strategic approach should not only attach great importance to technological innovation capabilities in commercial product development for the market (short term), improvement on product development (short term), and reformation of technological process for the market (short term), but also conduct more applied research for basic research and pay attention to basic scientific research.

3.4.4 Strategic Approach D: international R&D for international market

This approach for technological innovation is among the more advanced approaches for international technological innovation (Figure 3-11). With this strategic approach, enterprises usually carry out multiple sets of activities at the international level to achieve technological innovations. Each set of such innovative activities is aimed at the international markets and international technologies; the idea is to allocate and configure innovative resources globally and to gain capabilities globally. With this strategic approach, an enterprise should place its efforts on several important areas which include all the focal areas for the enterprises that were discussed earlier. Such areas include: extending and enhancing the enterprise's potential technological capabilities; providing necessary technical support for its products and technologies to its subsidiaries in its various local markets; bringing new products to those local markets as quickly as possible; making good use of good R&D hardware and infrastructure in the various local areas; making good use of any preferential policies in those areas; setting up technology monitoring points to know about its competitors and about the needs arising from new trend of technological development; recruiting local talent for R&D and absorbing advanced technology and scientific knowledge in those foreign areas; carrying out R&D projects in collaboration with various local companies or research institutions; and staying close to its suppliers.

Enterprises adopting this strategic approach are usually multinationals that are large in scale and powerful in strength. For such an enterprise, this approach for international technological innovation should include the following: establishing R&D branches in foreign areas, hiring researchers and carrying out R&D activities there, all through direct investment; acquiring foreign enterprises through capital market operations, with an effort to integrate the acquired technological capabilities and important resources for technological innovation into the enterprise; and, at the same time, engaging in collaborative research on a continued basis with foreign enterprises, R&D institutes, universities and so on to strengthen the technological areas where the enterprises are weak in terms of technological innovation. The R&D branches overseas will also work through various technology monitoring points to gain information about new developments overseas in relevant technology and knowledge areas.

With this strategic approach, there will be no central R&D locations, not even domestic ones. All R&D branches of an enterprise are connected and integrated into a relatively unified and harmonious network in which all knowledge and technology are shared. At the same time, each R&D branch is independent. In this way, each R&D branch can give full play to its initiative and enthusiasm and can better mobilize and energize its employees on both the technological and managerial sides of R&D and thus can improve the efficiency of its R&D activities. With this strategic approach, all of an enterprise's activities for technological innovation should form a comprehensive and integrated whole, and an enterprise's innovative activities should be balanced with both support for technological process and manufacturing as well as support for medium-term and long-term applied research and foundational research.

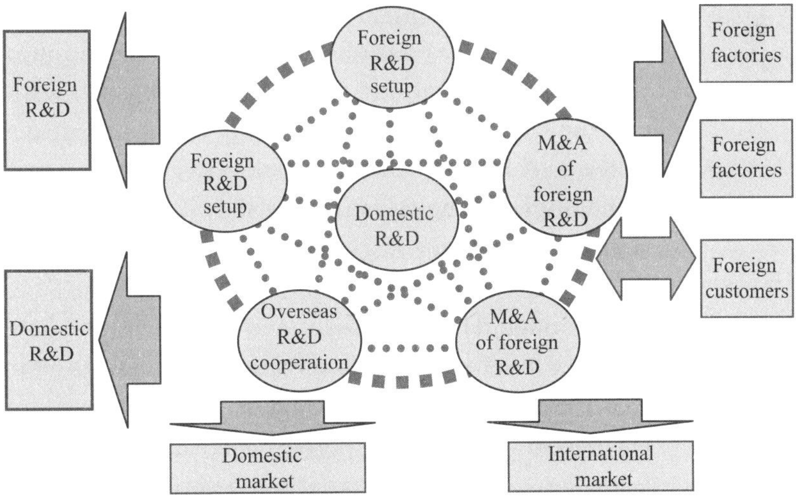

Figure 3-11 Strategic Approach D of internationalization of technological innovation for Chinese enterprises

3.5 Chapter summary

In this chapter we have analyzed and illustrated the important role of technological innovation in economic development and enterprise growth, the important role of internationalization of technological innovation in enterprise development in the ever more integrated world economy, and the necessity as well as the importance of the shift from the conventional strategy of enterprise technological innovation to one that is based on core competency.

Through detailed surveys and interviews, we have analyzed the basic situation of internationalization of technological innovation by Chinese enterprises, and, on the basis of this analysis, we have put forward our strategic framework for such internationalization by Chinese enterprises. The strategic framework we proposed is based on core competency, and it is a directional strategic framework intended for guiding Chinese business enterprises in the internationalization of their technological innovation activities.

We have summed up the four strategic approaches of internationalization of technological innovation for Chinese enterprises, which are as follows. Strategic Approach A: domestic R&D for domestic market with technology monitoring-oriented internationalization. Strategic Approach B: domestic R&D for domestic market first and then for international market with the purpose of international cooperation and technology monitoring. Strategic Approach C: domestic R&D branches mainly for domestic market, with overseas R&D branches established directly by the enterprise or obtained through M&As. Strategic Approach D: international R&D for international market. These four strategic approaches evolve from the lowest level to the highest level, and the attention is focused on the ongoing evolutionary process from domestic innovation to international innovation. In this chapter, we laid out both the preconditions required for adopting each of the strategic approaches and the priorities in the implementation process. The classification and the prescribed evolution processes of these strategic approaches are of practical significance in guiding Chinese enterprises to carry out internationalization of their technological innovation activities.

References

Ansoff, H. I. *Corporate Strategy*. New York: McGraw-Hill, 1965.

Barney, J. B. Firm resources and sustained competitive advantage. *Journal of Management*, 1991, 17(1): 99-120.

Cantwell, J. A. The globalization of technology: what remains of the product cycle model. *Cambridge Journal of Economics*, 1995, 19(1): 155-174.

Chandler, A. D. *Strategy and Structure*. Cambridge, MA: The MIT Press, 1962.

Chen, J. Learning mode from technology import to independent innovation. *Scientific Research Management*, 1994(15) : 32-34.

Chen, J. *Sustainable Development—New Perspective on Enterprise Technology Innovation*. Beijing: Science Press, 2001.

Chen, M. J. Competitor analysis and interfirm rivalry: toward a theoretical integration. *Academy of Management Review*, 1996, 21(1): 100-134.

Coombs, R. Core competencies and the strategic management of R&D. *R&D Management*, 1996, 26(4): 345-355.

Conner, K. A historical comparison on resource-based theory and five schools of thought within industrial organization economics: do we have a new theory of the firm here? *Journal of Management*, 1991, 17(1): 121-154.

Dalton, D. H. & Serapio, M. G. *Globalizing Industrial Research and Development*. Washington, DC: U.S. Department of Commerce, Technology Administration Office of Technology Policy, 1999.

Gao, J. & Fu, J. Y. Key issues in technology innovation for China's enterprises—1051 enterprises in technology innovation. *Technology Policy and Management in China and Abroad*, 1996(1): 24-33.

Gassmann, O. & Von Zedtwitz, M. New concepts and trends in international R&D organization. *Research Policy*, 1999, 28(2-3): 231-250.

Gerybadze, A. & Reger, G. Globalization of R&D: recent changes in the management of innovation in transnational corporations.

Research Policy, 1999, 28(2-3): 251-274.

Gimeno, J. & Woo, C. Y. Hypercompetition in a multimarket environment: the role of strategic similarity and multimarket contact in competitive de-escalation. *Organization Science*, 1996, 7(3): 322-341.

Hewitt, G. Research and development performed abroad by US manufacturing multinationals. *Kyklos*, 1980, 33(2):308-327.

Huang, H. X. *Market Innovation*. Beijing: Tsinghua University Press, 1997.

Liu, Y., Li, Y. & Yang, T. D. Analysis of impact on competitive advantages of technology innovation by state-owned industrial enterprises. *Economic Science*, 1997(1):19-23.

Mansfield, E., Teece, D. J. & Romeo, A. Overseas research and development by US-based firms. *Economica*, 1979(46): 187-196.

Meyer, M. H. & Utterback, J. M. The product family and the dynamics of core capability. *Sloan Management Review*, 1993, Spring: 29-47.

Mintzberg, H. & Quinn, J. B. *The Strategy Process: Concepts, Contexts, Cases*. London: Prentice Hall, 1992.

Niosi, J. & Godin, B. Canadian R&D abroad management practices. *Research Policy*, 1999, 28(2-3): 215-230.

Nonaka, I. & Takeuchi, H. *The Knowledge Creating Company: How Japanese Companies Create the Dynamics of Innovation*. New York: Oxford University Press, 1995.

Patel, P. & Vega, M. Patterns of internationalization of corporate technology: location vs. home country advantages. *Research Policy*, 1999, 28(2-3):145-155.

Patel, R. D. & Pavitt, K. Large firms in the production of world's technology: an important case of non-globalization. *Journal of International Business Study*, 1991, 22 (1): 1-22.

Porter, M. E. *Competitive Strategy*. New York: Free Press, 1980.

Porter, M. E. *Competitive Advantage*. New York: Free Press, 1985.

Prahalad, C. K. & Hamel, G. The core competence of the corporation. *Harvard Business Review*, 1992(March-April):79-90.

Robert, E. B. Benchmarking the strategic management of

technology (I). *Research Technology Management*, 1995a, 38(1): 44-56.

Robert, E. B. Benchmarking the strategic management of technology (II). *Research Technology Management*, 1995b, 38(2): 18-26.

Ronstadt, R. C. *R&D Abroad by US Multinationals*. New York: Praeger, 1977.

Ronstadt, R. C. R&D abroad by US multinationals. In: Stobaugh, R., Wells, L. T. Jr. (eds.). *Technology Crossing Borders*. Boston: Harvard Business School Press, 1984.

Rosenberg, R. *Inside the Black Box*. Cambridge: Cambridge University Press, 1982.

Schumpeter, J. *The Theory of Economic Development: An Inquiry into Profits, Capital, Credit, Interest, and the Business Cycle*. Beijing: Commercial Press, 1997.

Von Zedtwitz, M. & Gassmann, O. Market versus technology drive in R&D internationalization: four different patterns of managing research and development. *Research Policy*, 2002, 31(4): 569-588.

Wei, J. Core competency-based postacquisition integration management. *China Soft Science*, 2001(12): 62-65.

Wu, G. S. *Technology Innovation Management*. Beijing: Tsinghua University Press, 2000.

Xu, Q. R. *Research, Development, and Technology Innovation Management*. Beijing: Higher Education Press, 2000.

Zhao, Y., Cai, J. Y. & Luo, R. G. Technology innovation and model for construction of core competitiveness of China's enterprises. *China Soft Science*, 2001(1): 94-97.

Chapter 4

Internationalization of Technological Innovation: Organization and Management

4.1 Theoretical study on organization and management of internationalization of technological innovation

The synchronization between organizational form as well as processes and strategy is the prerequisite for effective execution of internationalization of technological innovation. With an increase in geographically dispersed internationalized technological innovation activities, effective organizational structures and management models for innovation activities have become the focus of attention of scholars and enterprises.

The quest by the management of many multinationals for a plan of innovation process management under the condition of internationalization of technological innovation oftentimes starts with a search for the correct organizational form. The multinationals believe that for each of the different technology strategies, there is always an ideal structure. When they get into more and more complex business environment in the internationalization process, they will naturally seek a new organizational model as the solution for the problem. However, changes in the organizational form themselves do not provide all the solutions due to the dynamic complexity faced by enterprises in internationalization of technological innovation activities. Organizational form is only a series of static roles, responsibilities, and relationships specified in the dynamic and rapidly developing target environment. Due to discrepancy in understanding between R&D branches and company's headquarters, their different opinions about information sharing, and the conflicts between demand for control and demand for autonomy, there are many contradictions in a global innovation network. Internationalization of technological innovation is not to solve a problem of structure, but rather it requires an organizational capability that is not easy to establish and operate. Enterprises need to balance cost, speed, and quality and thus are often faced with the dilemmas of centralization and decentralization, creativity and norms, face-to-face communication and telecommunication, and so on.

4.1.1 Organizational forms of internationalization of technological innovation

The organizational forms of internationalization of technological innovation are consistent with its organizational functions. Based on the organizational functions of internationalization of technological innovation activities, the famous German scholar Klaus Brockhoff (1998) grouped the organizational structures for such activities into three types.

(1) Star structure

This structure has the R&D center of the enterprise's headquarters in the center that coordinates with a number of overseas R&D institutions. In this organizational structure, the R&D center at the enterprise's headquarters is responsible for implementing innovation and the overseas R&D institutions are responsible for carrying out technology research, transferring the enterprise's technology overseas or developing products for the local market, and providing adaptive R&D and technical support for overseas subsidiaries; overseas R&D institutions are coordinated by that R&D center at the enterprise's headquarters and do not have much direct contact with each other.

(2) Multiple regional competency centers

This organizational structure has a central research institute in the enterprise's headquarters that coordinates with a number of overseas regional competency centers. Each of these regional competency centers consists of a number of R&D organizations, which coordinate with each other closely, develop products for the local market and the global market, or conduct certain basic research or basic applied research for the enterprise's long-term strategic considerations, with the goal of gaining proximity to knowledge centers. There are few links between the regional competency centers, and their activities are coordinated by the central research institute.

(3) Globally interconnected research organizations

This organizational structure is an organically coordinated global R&D network, which is composed of the central research institute and all the overseas R&D organizations. To serve the overall innovation strategy of the enterprise, this network utilizes global resources and carries out knowledge creation and technological innovation on a global scale.

In an empirical study, Bartlett and Ghoshal (1990) concluded that there are four types of universal overseas innovative organizations: local-for-global, local-for-local, global-for-global, and globally linked.

Gassmann and Von Zedtwitz (1999) summed up organizational forms for internationalization of technological innovation into the following five models.

(1) Ethnocentric centralized model

In this model, all R&D activities are carried out centrally in the home country. There is an asymmetry between the home country-based and other overseas R&D organizations with regard to information and decision-making, and the home country's technological innovation activities are strictly protected by the enterprise.

(2) Geocentric centralized model

When an enterprise becomes increasingly dependent on overseas markets and local capabilities, the ethnocentric centralized model will no longer be appropriate. Geocentric centralized model can help an enterprise to enhance its sensitivity to the local market and to reduce the cost of technological innovation for the local conditions. Obviously, this model requires that the enterprise increase its investment for R&D employees and raise their international awareness.

(3) Polycentric decentralized model

Each R&D branch has greater independence, and there are no special R&D centers above such R&D branches. However, in this model, the home country-based institution is often not able to obtain information about the activities of overseas R&D institutions; higher level of independence makes it impossible for them and other R&D branches to share information; and the organizational structure is overly bloated. Therefore, this organizational form is not preferred by most enterprises and researchers.

(4) R&D hub model

The home country's R&D center provides leadership for all the R&D works as well as R&D-related technological activities of the enterprise and thus reduces the risks of poor resource allocation as well as repetitive technological innovation projects. Overseas R&D institutions usually evolve from technological monitoring units that were previously set up in the home country. In this model, although the effectiveness of technology transfer can be ensured, the cost for coordination will be ever increasing; even worse, creativity may be inhibited and the flexibility and proactivity of R&D branches may be suppressed.

(5) Integrated R&D network

Under this model, the R&D center has evolved into a capability center, which is closely linked with each of the various independent R&D institutions through a variety of flexible and decentralized mechanisms. This network model is different from the R&D hub model, where each overseas R&D branch is an important component for the enterprise's strategy and the enterprise capability center must respond to the changes from various regions and must choose the right strategy and new business.

Actually, the organizational forms of internationalization of technological innovation are not unalterable; instead, they are in constant change. For example, to improve the efficiency of internationalization of technological innovation, many enterprises have gradually integrated their technological innovation activities into a small number of technology centers of excellence; to meet the needs of international market, R&D organizations have become ever more open and the process management of technological innovation is being adapted to the international market as well as knowledge centers and is evolving in the direction of an integrated innovation network.

4.1.2 Centralization and decentralization

When formulating strategic decisions, an enterprise will face the question: how much autonomy should be granted to the local R&D institutions? Decentralized resources and decision-making can improve the ability to rapidly respond to the overseas market, but the lagging effect of knowledge dispersion and flow will affect the effectiveness of organizational learning and will cause efficiency loss. Meanwhile, too much autonomy will lead to conflicts and difficulty in controlling the local R&D institutions. The concentration of resources and capabilities at the headquarters can help obtain efficiency, leading to a rapid creation of new products and technological processes; however, excessive control will stifle creativity, prohibiting the full use of local capabilities.

From the point of view of the market, internationalization of technological innovation also faces contradictory choices: is it better to centralize technological innovation in one area in order to improve efficiency or to decentralize innovation so as to get closer to customers? The central innovation team members usually lack actual understanding of the needs of foreign markets and production, and although product development for specific market segments and even specific customers can increase sales, it will also increase costs. In the past, most enterprise management chose "compromise," but the cost of doing so is to lose the creative tension that is very important for innovation activities.

Behrman and Fischer (1980) summed up five models for decision-making in a global innovation network: (i) absolute centralization—all decisions are made by the company's headquarters, and subsidiary units are only responsible for the implementation; (ii) participatory centralization—subsidiary units participate in decision-making; (iii) cooperative decision-making—subsidiary units and the company's headquarters come together on an equal footing to make decisions by agreement; (iv) supervised liberalization—decisions are mainly made by the subsidiaries under the supervision of the company's headquarters; and (v) full liberalization—each subsidiary acts with full autonomy.

Arimura (1997) studied from another angle and summarized various forms of decision-making into three types: top-down approach (absolute centralization), bottom-up approach (full liberalization), and mixed approach. In his view, what form of decision-making to use is itself a decision to be made on the basis of the nature of the project: for strategic projects, decisions should generally be made by the company's headquarters or under the supervision of the company's headquarters;

for regional or local projects, decisions should be made by the subsidiary units.

Brockhoff (1998) defined 12 decision-making activities related to the degree of coordination at the laboratory level, including formulating innovation plan; initiating cooperation; holding project selection; planning strategy meeting; holding operations meeting; and deciding project budget, cost-minimizing strategies, and personnel arrangement, as well as four decision-making units—central R&D unit, other central units, local R&D unit, and other local units; and with such definitions, he systematically studied the relationship between centralization and decentralization. In his view, whether or not the coordination within a global innovation network will be successful depends on a combination of three variables—the degree of centralization for decision-making, the degree of autonomy for the laboratories, and the capabilities of the laboratories.

4.1.3 Exchange and communication

Effective use, transmission, exchange, and communication form the foundation for successfully implementing internationalization of technological innovation. However, because multinationals have expanded their technological activities to the entire globe, effective use and transmission of information and knowledge have become difficult. Thus, studies on communication within an international innovation team among members from different companies, different countries, and different cultural backgrounds are absolutely necessary. Internationalization of technological innovation has resulted in the distribution of R&D branches on an international scale. The dispersion of R&D activities, internationalization of R&D employees, increase in the number of external partners, complexity of the network, and cultural and linguistic differences have brought significant challenges to exchange and communication. In order to promote effective communication in the process of technological innovation,

R. K. Moenaert, F. Caeldries, A. Lievens, and E. Wauters (2000) conducted systematic studies on communication in the process of internationalization of technological innovation. In their view, an innovation team is in itself an information processing system and an organizational structure capable of successfully managing an internationalized innovation project is a structure that can match information processing capabilities with information processing needs. Therefore, there are five necessary conditions for effective communication in the process of internationalization of technological innovation as follows.

(1) Transparency of the communication network

If the network is not transparent, members of the network will have problems in identifying information source and determining the direction of information flow. However, because the initial stage of an international innovation project is often filled with uncertainties, it is difficult to construct a transparent communication network.

(2) Codification of knowledge

New product development in an international environment may stop the knowledge codification process. First, different units and functional departments of the organization will form their own special "languages," subcultures, and methods to explore and analyze the environment. Second, the linguistic, cultural, and environmental differences among the countries make encoding more difficult.

(3) Credibility of knowledge

In communication among the geographically dispersed parties, many pieces of important social/situational information—such as a person's social status and level of technological expertise—may go missing or get distorted. This may change or hinder the process of trust building among team members; as a result, the cross-functional interfaces will likely be filled with an atmosphere of distrust and this situation will have a great negative impact on the communication among various functional departments, reduce the credibility of information, and lower information acceptance level.

(4) Cost of communication

Distance is an obvious obstacle to interpersonal communication. Whether it is frequent and direct communication or communication through modern telecommunications technology, its time and financial costs are always an issue. Cost restrictions pose tremendous resource uncertainties to geographically dispersed organizations.

(5) Confidentiality

When technological innovation activities are concentrated in only one place, it is easy to ensure confidentiality. However, when an enterprise carries out trans-regional technological activities, it has to take every possible step to prevent leakage of secret information. The internationalization of technological activities increases the possibility of leaking secret information, and the need for confidentiality to a certain extent hinders communication.

Besides, in order to achieve effective communication, organizations must develop capabilities at the enterprise level as well as at the team level.

Enterprise-level capabilities include the following: (i) forming informal networks, setting up an international steering committee, and letting them operate in parallel with the formal hierarchical structure, so as to address the contradiction between autonomy and control; (ii) creating an atmosphere of cross-functional coordination, so as to improve cohesion; (iii) raising the level of communication infrastructure, so as to increase the opportunity for information exchange; and (iv) establishing a consistent long-term goal, so as to prevent the situation where each branch will focus on its local operations and short-term projects and not be willing to bear the risk of long-term technological innovation activities.

Team-level capabilities include the following: (i) establishing the core team, which will be responsible for managing and coordinating cross-functional projects; (ii) correctly selecting team leaders, so as to reduce the cost of communication and increase the degree of information security, network transparency, and knowledge credibility; (iii) formalizing teams—while an informal work environment and a certain degree of autonomy are very important for product development, the functions, responsibilities, and key points of the entire project must be clearly laid out; and (iv) incorporating various functional departments into project development as early as possible, which is crucial to the introduction of new ideas into a successful commercialization process.

4.1.4 Integration of internationalization of technological innovation

Many companies strive to achieve a situation where the assets and resources, despite being widely distributed, support each other to achieve effectiveness and efficiency on a global scale. The functions and duties of each branch are not the same, but the branches should be complementary to each other, so as to promote learning on a global scale, facilitate company-wide knowledge sharing, and ensure the effective development and dissemination of innovation. In such an organization, the headquarters and branch offices have clear responsibilities, but at the same time, the management processes connecting various units become more complex.

Business enterprises have been continuously working to integrate the chains of innovation process. The past questionable practice to isolate basic research, applied research, development, products, and practical needs should be corrected. Multinationals and leading research centers continue to improve their competitiveness by virtue of the interactions among research, advanced manufacturing, and advanced market. The integration of product development process and related technological processes, and the cross-border integration of R&D, marketing, production, and products have become the key issues in innovation management (cf. Leonard-Barton, 1995; Nonaka & Takeuchi, 1995).

By summarizing previous studies, we can divide process integration management of internationalization of technological innovation into the following two types. (i) Complementary process integration—this type of management tool is complementary to organizational structure or procedures (see Table 4-1). It can be about personalized activity (developing common vision and strategy for the organization and learning and understanding the various activities) and impersonalized activity as well (effective reporting rules and standards, concurrent engineering, job rotation, and design of development plan for a collaborative corporate culture). (ii) Process integration with someone working full time on it—the managers can be appointed as project leaders, process advocates, coordinators and contact persons, and can also occur naturally for the needs in the work process, such as internal star staff, cross-border personnel, etc.

Von Zedtwitz (2002) found that the technological innovation process can be divided into the following three phases, which can expedite the innovation progress. (i) Early phase of project. In the early phase of technological innovation process, the process flow is nonlinear and smart thinking must be formulated. There are no geographic limitations as to where to look for the knowledge of new technologies; however, invention and innovation are usually regional activities. System architecture needs to be clearly defined. (ii) Project implementation. Project development must be strictly and efficiently implemented. During this phase, global team cooperation is possible, but responsibilities and interfaces need to be clearly defined. (iii) Market introduction. There may be different strategies for this phase: introduction of a new product to market can happen in one market at a time or with coordination in all the markets at the same time.

Table 4-1 Ways of complementary process integration

Positioning	Personalized	Impersonalized
Structure	Team building —Committee; —New product group; —Project team; —Task group; Matrix organization; Willingness to integrate different cultures; Consideration for cooperation and support in personnel selection	Distance reduction through the following means —Decentralization; —Space arrangement; Planning with decomposition rules; Documentation of plan; Definition of cross-unit joint product platform; Coordination by price mechanism; Data bank on projects and equipment
Process	Common goals and standards; Development of common vision and strategy for the organization; Avoiding extreme solutions for local or functional goals; Use of motivational systems for international coordination and co-operation; Learning and understanding different activities by participating in different jobs or receiving more education; Internal information meeting and support for personal contact	Updating information about plan through information exchange; Effective reporting rules and standards, such as technological development road map; Network work as a turning point; Concurrent engineering; Job rotation; Design of development plan for a collaborative corporate culture

Source: Brockhoff, K. *Internationalization of Research and Development*, Berlin: Springer-Verlag, 1998.

4.2 Analysis of the organization and management of internationalization of technological innovation for Chinese enterprises

Among the received filled-in questionnaires, not all of them have answers for every question about organization and management of internationalization of technological innovation. The questionnaires that do not have answers for the questions pertaining to organization and management are not considered for the analysis here.

4.2.1 Organizational form

As depicted in Figure 4-1, Chinese enterprises mainly rely on the domestic R&D institutions to conduct innovative analysis and implement its results, that is, to discover new opportunities in the domestic market, apply the parent company's R&D capabilities to develop new products and new technological processes, and then put them to use in appropriate locations around the globe. Overseas R&D units basically exist as technology monitoring points and thus do not undertake R&D projects that will have a major impact on the company. Even if an overseas branch is to carry out a major R&D project, the project will be under the strict control of the R&D unit in the company's headquarters. Decisions about the R&D project, budget approval, and so on will all be made by the company's headquarters, as overseas branches do not have much autonomy in this regard. We believe that such an emphasis on the control of overseas branches is in line with the size and strength of a typical Chinese enterprise with its present operating situation. Presently, Chinese enterprises have limited strength and are not able to establish R&D branches distributed worldwide, and they even have difficulty in setting up regional R&D centers in certain areas of the world.

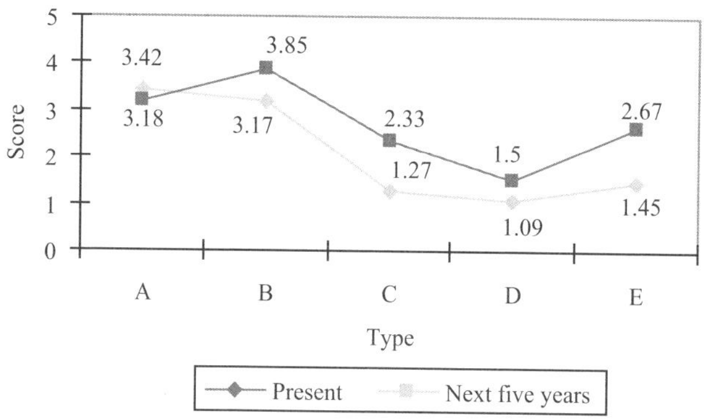

A: R&D activities are carried out mainly in the home country, and little attention is paid to overseas R&D;

B: R&D headquarters is set up in the home country, and the headquarters directs overseas R&D activities and exerts strong control over overseas R&D activities;

C: Regional R&D centers are set up in parallel in a number of regions of the world, and such centers guide their subordinates to carry out R&D activities;

D: Independent R&D institutions are set up around the world, and the headquarters exerts relatively weak control over them;

E: A closely interconnected worldwide R&D network is constructed, and this network enhances the exchange and communication as well as information transfer among the R&D institutions that are dispersed around the world.

Figure 4-1 Organizational forms of overseas R&D branches of Chinese enterprises

However, when it comes to expectations of the surveyed enterprises, such situations will change. With the development of its internationalized operations, an enterprise will find different needs and opportunities in various countries: consumption is more complex in certain areas, competition is more intense in some other areas, technology used is worth more attention in some other areas, and products developed by the home country-based central research institutes are not guaranteed to succeed abroad. At the same time, with the emerging new international opportunities after China's accession to the WTO, Chinese enterprises will begin to rely more and more on the use of resources from both central research institutes and overseas branches for solutions. Therefore, the enterprises will pay more attention to adapting their products for overseas markets, and thus, the overseas R&D institutions will become increasingly important. In the process of setting up overseas R&D institutions, the enterprises will still attach great importance to the control and guidance of overseas branches; such organizational structure based on the home country and supplemented by the overseas R&D activities will become the most important structural form of internationalization of technological innovation for Chinese enterprises in the next five years. Obviously, limited by the strength of Chinese enterprises, the central research institute's focus on product and process design work is yet another important aspect for managing the technological activities in these enterprises. In addition, globally interconnected R&D network will become a trend that cannot be ignored. As for the innovation organizational form where there are multiple independent centers and the headquarters exerts relatively weak control, it is still not valued by business enterprises, which confirms the conclusion drawn by Gassmann and Von Zedtwitz in 1999.

Further analysis on what industry each of the surveyed enterprise belongs to showed that in the internationalization process of technological innovation, not many differences between the organizational structure taken by each industry and that taken by all the enterprises as a whole have been noticed. Under the current situation in every industry, companies attach great importance to the strength of control by company's headquarters and prefer centralized technological innovation mode and organizational structure (see Figure 4-2). Compared with the traditional mechanical and electrical appliance enterprises, the high-tech enterprise need to have a better understanding of the trends in the S&T development worldwide and integrate domestic and overseas R&D resources in order to provide solutions. However, after all, the high-tech industry is an emerging sector; most domestic high-tech enterprises are not strong, and thus, it is not realistic to expect them to set up R&D institutions worldwide. In the expectations of the surveyed enterprises for the next five years (see Figure 4-3), on the one hand, the star organizational structure is still being highly recognized by the investigated enterprises and the domestic R&D activities are also considered important for an enterprise's technological innovation. On the other hand, to the other three organizational forms, each industry attaches different levels of importance. Large machinery enterprises attach greater importance to geocentric centralized organizational structure, and the score given by them for this structure increased from the original value of 1.2 to 2.8; high-tech enterprises are more likely to see globally interconnected organizational structure as the direction of future development, and the score given by them for this structure increased from the original value of 1.75 to 2.8; and enterprises in the home appliance industry are also more likely to choose the globally interconnected innovation mode with an increased score from 1.0 to 3.0.

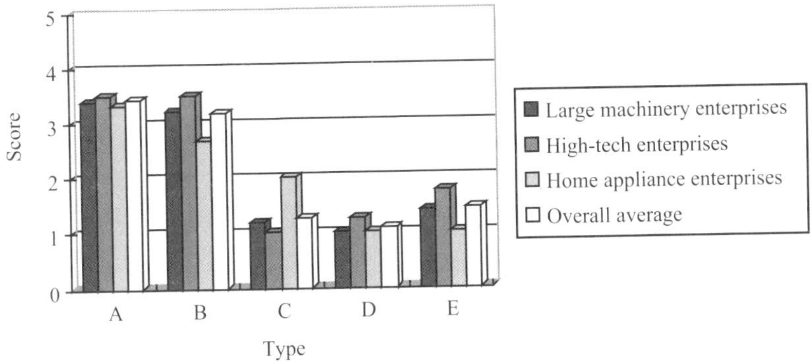

A: R&D activities are carried out mainly in the home country, and little attention is paid to overseas R&D;

B: R&D headquarters is set up in the home country, and the headquarters directs overseas R&D activities and exerts strong control over overseas R&D activities;

C: Regional R&D centers are set up in parallel in a number of regions of the world, and such centers guide their subordinates to carry out R&D activities;

D: Independent R&D institutions are set up around the world, and the headquarters exerts relatively weak control over them;

E: A closely interconnected worldwide R&D network is constructed, and this network enhances the exchange and communication as well as information transfer among the R&D institutions that are dispersed around the world.

Figure 4-2 Cross-industry comparison of organizational structures in the section on internationalization of technological innovation for Chinese enterprises

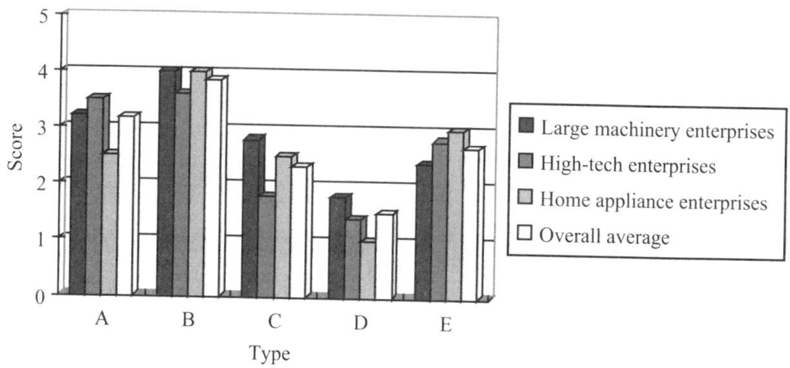

A: R&D activities are carried out mainly in the home country, and little attention is paid to overseas R&D;

B: R&D headquarters is set up in the home country, and the headquarters directs overseas R&D activities and exerts strong control over overseas R&D activities;

C: Regional R&D centers are set up in parallel in a number of regions of the world, and such centers guide their subordinates to carry out R&D activities;

D: Independent R&D institutions are set up around the world, and the headquarters exerts relatively weak control over them;

E: A closely interconnected worldwide R&D network is constructed, and this network enhances the exchange and communication as well as information transfer among the R&D institutions that are dispersed around the world.

Figure 4-3 Cross-industry comparison of organizational structures for the next five years in the section on internationalization of technological innovation for Chinese enterprises

Thus it can be seen that after China's accession to the WTO, Chinese enterprises have been facing a variety of opportunities and challenges and their choices of the organizational modes for internationalization of technological innovation have also become diversified. The difficulties of Chinese enterprises in organizational restructuring for internationalization of technological innovation are as follows: because of the source of their traditional competitive advantages, the point of view of the enterprises can be biased; some enterprises attained competitive advantages through the economy of scale realized in concentrated production, and internationalization of technological innovation based on geographical dispersion obviously does not conform to this traditional organizational form, and thus, such enterprises are still biased toward organizational forms for technological innovation with higher level of control. In the process of establishing a new, more complex, and dynamic operating mode for internationalization of technological innovation, Chinese enterprises must do away with such biases.

4.2.2 Exchange and communication

The notable features of internationalization of technological innovation are the globally dispersed R&D branches and the internationalized employees. One early study indicated that the frequency of face-to-face communication between the central laboratory and an overseas subsidiary goes down exponentially with the geographic distance between them (Hough, 1972). For this reason, many scholars have suggested that companies need to make use of various approaches, unprecedented information resources, and information technologies to provide new opportunities for managing R&D activities. A company can "shorten" the geographic distance by using modern electronic communication technology and integrate the dispersed technological innovation activities with high-efficiency technological platform; in this way, the company's headquarters and its various R&D branches can form a closely linked network, so as to promote the company's cultural identity among R&D employees, promote the understanding of its strategic intent by the R&D employees, and eventually improve the effectiveness of its technological innovation activities (Gerybadze & Reger, 1999; Von Zedtwitz, 2002; Gassmann & Von Zedtwitz, 1999; Golder, 2000).

Although not many Chinese enterprises have set up R&D institutions abroad, the fact that technological innovation activities are carried out internationally still has greatly changed the way technological innovation activities are managed. Our result shows that the surveyed enterprises generally value and stress the use of information technology for communication and that the mean of the scores given to this area is 3.73. E-mails, video technology, electronic conferences, groupware, and so on have been widely used by Chinese multinationals, but IT tools are not as effective as direct face-to-face interaction, and communication through modern telecommunication tools are still not entirely satisfactory. Due to the emphasis on control by the headquarters, enterprises attach greater importance to communication between the headquarters and the branches than those among the branches themselves and, as a result, the communication between the branches and the headquarters are better than those among the branches themselves. The score assigned for communication effectiveness between the parent company and the branches is 2.8 and that among the branches is 2.22 (see Figure 4-4). Relatively speaking, high-tech enterprises are better in the application of IT tools for communication (the score given for the use of IT tools between the headquarters and the branches is 4.25 and among the branches themselves is 3.5) and they attach greater importance to face-to-face communication among the overseas branches (scored at 2.25). However, traditional home appliance enterprises and machinery enterprises hold similar views on communication, and the differences in assigned score for four of the six factors are under 0.1. It should be noted that face-to-face communication among the branches are almost never considered by the home appliance enterprises and large machinery enterprises (with an assigned score of 1.0 for both types of enterprises).

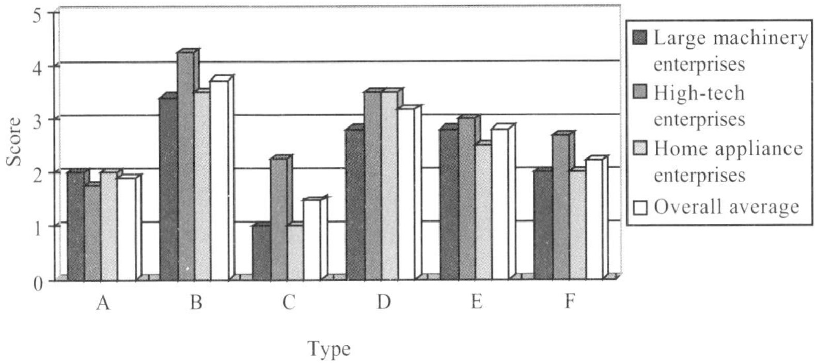

A: Parent company and its overseas R & D branches mainly communicate via face-to-face contact;

B: Parent company and its overseas R & D branches mainly communicate through IT tools;

C: The overseas R & D branches mainly communicate with each other via face-to-face contact;

D: The overseas R & D branches mainly communicate with each other through IT tools;

E: Present level of satisfaction with communication between parent company and its overseas R&D branches;

F: Present level of satisfaction with communication among its overseas R&D branches.

Figure 4-4 Exchange and communication involving overseas R&D branches

With the spread and popularization of IT tools, each technological innovation center no longer needs to send people back and forth frequently for contacting other centers; also, IT tools are gradually taking the place of face-to-face communication. Although IT tools are the only necessary means, they are not enough to eliminate the regional differences and time differences, not to mention the challenges of cultural differences. The basis of mutual learning between two individuals is trust, but a transnational interface for technological innovation is often filled with an atmosphere of distrust because of the low frequency of communication and the differences in message interpretations; such situation can often have a great negative impact on the communication between the functional departments (Moenaert et al., 2000).

From this point of view, just like foreign enterprises, Chinese enterprises also need to find the balance point between face-to-face communication and IT-assisted communication when carrying out transnational innovation, and enhancing communication among the overseas branches as well as between the overseas branches and the R&D center of the headquarters should become an important specialization in international innovation management.

Considering the degree of importance for each of the ways to create organizational openness, the surveyed enterprises attach particular importance to exchange and communication in the process of implementing internationalization of technological innovation between the person in charge of an overseas branch and the headquarters. The overall rating for this factor is 3.93; the goal of such communication is to ensure that the R&D branch fully understands the business strategies and intentions of the enterprise. The surveyed enterprises value regular contact among the managers and R&D personnel of overseas R&D branches (scored at 3.43) and encourage the branches to establish a culture of open information exchange (scored at 3.64). The overseas R&D branches encourage holding various formal and informal seminars (scored at 3.14) and organizing international teams and also encourage researchers from different countries to establish personal contact and to influence each other (scored at 3.36). However, the surveyed enterprises do not pay enough attention to the rotation of jobs at different branches for employees and to introducing employees to each other of the cultural and social background and habits in their respective countries. For language training and introducing employees to each other of the cultural and social background and habits in different countries, both industry average and overall average are under the score of 4. Under the current situation where internationalization of technological innovation has not yet reached maturity, lagging behind on this count will have a negative influence on the exchange and communication among the employees of different nationalities (see Figure 4-5).

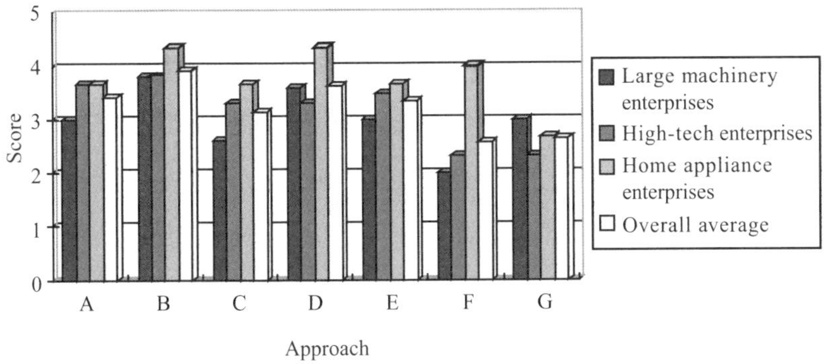

A: To encourage managers and R&D personnel of overseas R&D branches to be in regular contact;

B: To encourage managers and management personnel to exchange and communicate regularly with R&D headquarters of the parent company;

C: To hold formal and informal seminars regularly at overseas R&D branches;

D: To create a culture of open information exchange at overseas R&D branches;

E: To organize an international team and encourage researchers from different countries to establish personal relationships among themselves and to influence each other;

F: To rotate jobs at different overseas branches for researchers and management personnel;

G: To organize internal language training at overseas branches, where participants introduce to each other the cultural and social background, habits, and so on of different countries.

Figure 4-5 Various approaches to creating organizational openness at an overseas R&D branch

The home appliance enterprises are most active in this aspect among the various industries—they attach great importance to the communication between the overseas branches and the enterprise's headquarters (scored at 4.33), creating a culture of open information exchange (scored at 4.33), and striving to facilitate the rotation of jobs in different overseas branches for employees (scored at 4.0). Large machinery enterprises are not very active in this aspect, and except for in the area of culture building ("to create a culture of open information exchange" at 3.46 and "to organize language trainings... introduce to each other the cultural as well as social background, habits, and so on of different countries" at 3.0), they do not perform as effectively as high-tech enterprises and home appliance enterprises. Machinery industry is the most traditional industry in our survey and vertical communication in a hierarchical system is the main communication approach in the traditional management mode, but many cross-border innovation projects attach great importance to horizontal communication; thus, it can be imagined that machinery enterprises will be faced with greater difficulties than those of the other two industries in the effort to achieve good communication in a geographically dispersed technological innovation process.

In the collected questionnaires, factors depicted in Figure 4-6 are all considered as important ones in contributing to innovation project failures. Among all the factors, the risk of innovation project failure as a result of fund shortage is considered smaller by the enterprises than that as a result of any of the other five factors and the overall score for this risk is 2.5, which indicates that the surveyed enterprises have a good understanding of the importance of technological innovation and thus innovation projects can get the enterprises' financial support. The overall scores for the risks of the other five factors are all above 3, with few differences among them.

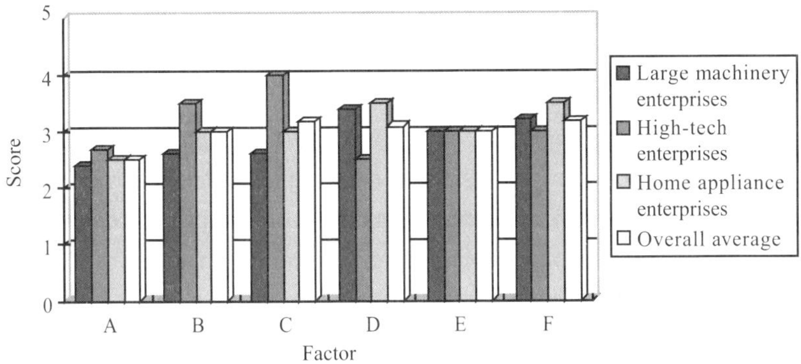

A: Fund shortage;

B: Degree of technological difficulties;

C: The level of technological content and level of commercialization of the enterprise's product are lower than those of its competitors;

D: Difficult to control the complex relationships with its overseas partners;

E: Cultural differences, language differences, and habit differences between Chinese and foreign researchers lead to difficulties in exchange and communication, prevent the formation of effective cooperation and eventually make it difficult for research projects to be smoothly implemented;

F: Domestic parent company and overseas R&D branches cannot carry out effective exchange and communication, and as a result overseas R&D branches cannot fully implement the technological strategy and technological goals of the parent company.

Figure 4-6 Factors for failures of technological innovation projects at overseas R&D branches

The situation in each industry is slightly different from that of the industries as a whole. High-tech enterprises believe that the failure of a cross-border innovation project can most likely be attributed to its technology aspect: difficulty in achieving the technology solution (scored at 3.5) or lower level of technological content and lower level of commercialization of its own product than those of its competitors (scored at 4.0). They talk less about "soft" factors such as culture building and exchange as well as communication as being the reasons for failure. The technologies applied by high-tech enterprises are the relatively frontier technologies of the world; such technologies themselves are still not fully developed, and thus high-tech enterprises bear greater technology risks when carrying out innovation activities. However, large machinery enterprises and home appliance enterprises believe that relationship with partners, cultural differences within R&D institutions, and communication problems are the more important reasons for the failure of international innovation projects; the scores they gave for these factors are above 3.

The ability of one laboratory by itself is limited; thus, technological innovation under many situations are carried out in cooperation with an overseas R&D institution, or for some innovation projects, multiple R&D institutions worldwide work together on development. This approach can not only reduce the risks of local R&D but also make use of the benefits of global learning. However, the coordination cost of this approach is high, especially when the coordination happens within a cross-cultural environment for the relationships with overseas partners and for achieving effective information exchange. Therefore, problems with culture and communication have become important factors for the failure of international innovation projects.

4.2.3 Integration of internationalized innovation processes

Since the mid-1990s, multinationals have begun integrating their business processes and improving their organizational efficiency. Decentralized technological activities and globalized innovation processes have greatly increased the structural complexities of business organizations and have made them difficult to manage. Thus, it has become necessary for such enterprises to seek a more streamlined and more effective organizational form for the management of their global technological innovation activities. For Chinese enterprises in the process of cross-border innovation, process integration will have a direct impact on the outcome of internationalization of technological innovation.

Surveyed enterprises roughly gave the same responses to the factors depicted in Figure 4-7 about their individual impact on the outcomes of major international innovation projects involving multiple overseas branches. Setting up a specialized team in coordinating various R&D units (overall score 3.64) and ensuring the goal congruence of the various units (overall score 3.79) are the most important factors to the surveyed enterprises. Only when the goals of the various units are congruent, can it be possible that the innovation processes at various units proceed in sync. In the case where a number of units jointly develop a large-scale innovation project, the responsibility of each unit must be clearly defined; thus, the project should be divided into a number of subprojects, each of which should be assigned to an R&D unit; meanwhile, to avoid the loss of team enthusiasm due to the prolonged time period of a large-scale project, it is necessary to divide the implementation of an innovation project into key phases, which can be used for evaluating performance in the innovation process. To ensure that different units proceed in the same pace, it is necessary to let the various units know each other's progress. The surveyed enterprises also believed that in order to promote the transfer of tacit knowledge, the system should be built in such a way as to ensure face-to-face communication and staff mobility.

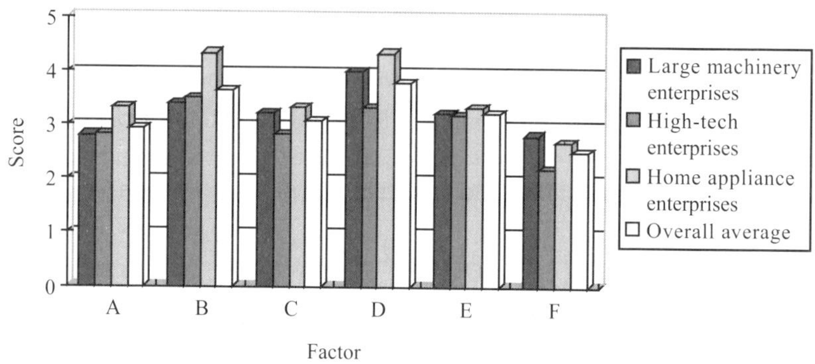

A: Division of a large project into subprojects and its implementation into key phases;

B: Setting up a team specialized in coordinating the various R&D institutions;

C: Regular information exchange among R&D institutions;

D: Goal congruence of the various R&D institutions;

E: Progress synchronization and uniformity of various R&D institutions in accomplishing their assigned tasks for each phase—research, development, integration, testing, etc.;

F: Moving R&D personnel along when carrying out knowledge transfer among R&D institutions.

Figure 4-7 Factors affecting the outcomes of major innovation projects involving multiple overseas R&D branches

Home appliance enterprises respond more positively to such factors. For R&D personnel mobilization, their score is 2.67, which is higher than that of other industries except for the machinery industry. Their scores for other factors are the highest among those of all the industries. The factors they value the most are also setting up a coordination team and ensuring goal congruence, with both scores being above 4.33. As for high-tech enterprises, they are better at using IT tools for communication than home appliance and large-scale machinery enterprises; thus, they do not attach as much importance to moving

R&D personnel around as the means for accomplishing knowledge transfer as enterprises in those two industries do, and they gave that factor a score of only 2.17, but other factors all have a score above 2.8. The scores given by the large machinery enterprises for these factors do not differ much from each other, except for "ensuring goal congruence" which has a score of 4.0, all the other factors are given a score between 2.8 and 3.2.

Von Zedtwitz (2002) found that the technological innovation process can be divided into three phases: (i) early phase of project; (ii) project implementation; and (iii) market introduction. Meanwhile, Moenaert et al. (2000) concluded that acceptance and faithful implementation of a decision to a large extent depend on the early stage involvement of various departments that will be affected by the decision. Thus, involving the various functional departments as early as possible in the project development is very important for introducing new ideas into a successful commercialization process.

When judged from the situation of the surveyed enterprises, Chinese enterprises are not fully aware of involving various departments in the development process. As shown in Figure 4-8, whether at the industry level or for all the surveyed enterprises as a whole, the average scores for this item are always around 3.0, which indicates that these enterprises do not really have a unified understanding of when non-R&D departments should be involved in innovation projects. Closer analysis of the collected questionnaires revealed that the number of enterprises that involve non-R&D departments in innovation projects during each of the three phases is also roughly the same. Large machinery enterprises and home appliance enterprises are slightly inclined toward involving non-R&D departments during the market introduction phase of innovation projects, but the surveyed high-tech enterprises are almost indifferent to during which phase to involve non-R&D departments in innovation projects.

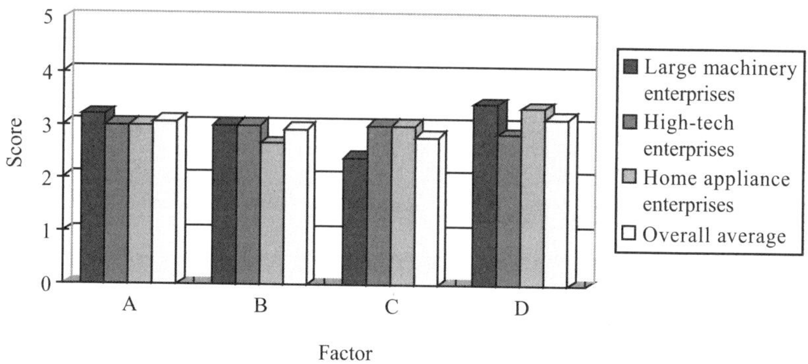

A: Involvement of non-R&D departments at the start of project planning phase;

B: Involvement of non-R&D departments during the early stage of project implementation;

C: Involvement of non-R&D departments during the middle or late stage of project implementation;

D: Involvement of non-R&D departments during the market introduction phase.

Figure 4-8 Timing factor of non-R&D departments involvement in innovation projects

When carrying out an internationalized innovation project, given that the various departments involved are geographically dispersed and since existing international communication networks are not transparent, the problem mentioned in the previous paragraph is more acute. After joining the WTO, Chinese enterprises are facing greater market opportunities and challenges. If the various functional departments do not get involved during the product description phase and the strategic planning phase, R&D department will not get all the needed information and non-R&D departments cannot be guaranteed to be motivated during the middle and late stages of the project. As a result, the project will quite likely run into problems after product development and project launch.

When an innovation project has succeeded technologically, the innovation process does not truly end; it has to go through the commercialization stage, which is to introduce the fruit of the innovation into the market; and thus, the timing for introducing the new product or service becomes an important factor in determining whether or not an innovation project will succeed—being too late or too early will make it impossible to achieve the maximum benefits. If the introduction happens too late, the product will be faced with fierce competition, potential users may have already accepted the style or structure of competing products, and profit margin will thus be greatly reduced; if the introduction happens too early, project failure may also happen because potential users still do not know the need for the product and the enterprise cannot profit from product support services.

On the whole, there are no fixed patterns for the choices made by the surveyed enterprises for the timing of market introduction (Figure 4-9). To a slight degree, more enterprises choose to carry out product commercialization after the market has already been activated initially, with an average score for this factor above 3. Considering the product differences among the industries, we believe that this situation is normal. Home appliance enterprises are more likely to enter a market after it has already been activated initially, and the cases with products such as VCD, DVD, rear projection TV, and so on in the home appliance market of China have confirmed this point. High-tech enterprises are less likely to enter a market after it has matured and thus they give this factor a score of only 1.8. The choices made by large machinery enterprises are rather spread out, and their scores for the three items shown in Figure 4-9 are all between 3.2 and 3.4.

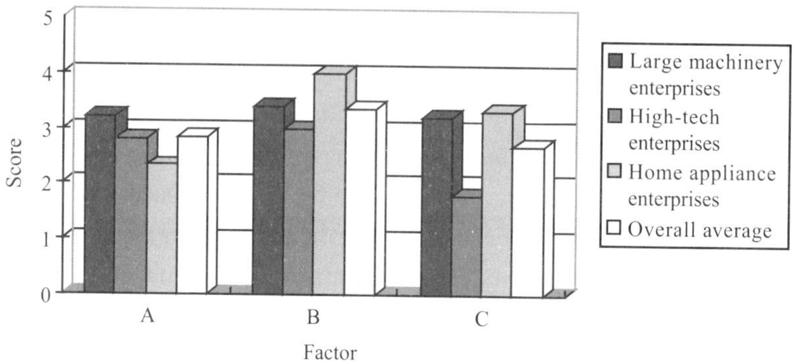

A: Innovation projects for stimulating consumer demand (that is, for opening up new market);

B: Innovation projects to meet consumer demand that has already been activated;

C: Innovation projects for a mature market.

Figure 4-9 Timing factor for introducing innovative products into the market

For making an entry into the overseas market (Figure 4-10), product export is still the main approach for Chinese enterprises, especially for home appliance enterprises, which gave this item an average score of 4.5. Aside from that, setting up wholly owned subsidiaries and establishing joint ventures with the local companies are also the main approaches for Chinese enterprises to enter the overseas market. Meanwhile, few of the surveyed enterprises enter the overseas market by controlling the local companies through M&As; we believe that this is due to a lack of financial strength—aside from a small number of very strong enterprise groups, most Chinese enterprises do not have enough financial strength to acquire companies in the overseas market yet. It is worth noting that home appliance enterprises hardly consider the OEM (Original Equipment Manufacturer) approach (giving it a score of 1). High-tech enterprises do not value any approach in particular, but obviously, they pay more attention to the rights to use patents than those in the other two industries.

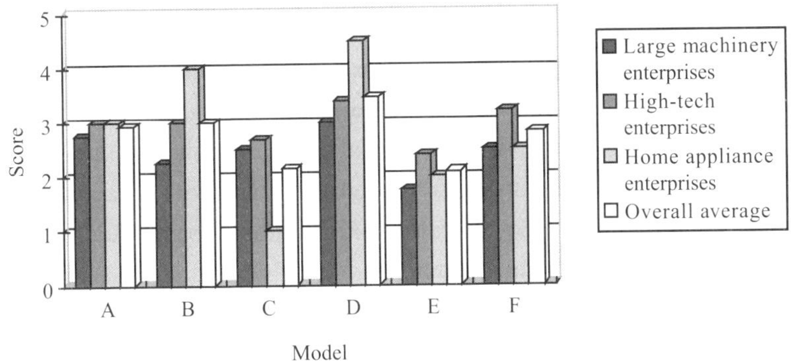

A: Wholly owned subsidiary;

B: Joint venture;

C: OEM (outsourcing);

D: Product export;

E: M&A (controlling a company in an important market by means of M&A and making use of the fruit of the innovation projects and carrying out production, sales, and so on through this company);

F: Transfer of the rights to use a patent.

Figure 4-10 Models for introducing innovative products into the overseas market

To the methods of control listed in Figure 4-11, the surveyed enterprises mostly responded in consistent ways; most enterprises adopt more than one approach to control the overseas R&D institutions. Among them, more high-tech enterprises and home appliance enterprises recognize these methods of control and the average scores given by them reached 3 or above, but large machinery enterprises probably adopt a different method of control and they gave scores between 2 and 3 for the listed six methods. However, we have found through the survey that there is a problem with the innovation process control adopted by these enterprises: in many cases, the innovation process is controlled as a common process (such as a production process) (see Item C in Figure 4-11, with an overall average score of

2.9) and tasks at various R&D institutions are standardized as much as possible (see Item E in Figure 4-11, with an overall average score of 3.1). However, in fact, the process of internationalized innovation is filled with uncertainties, and the organizational design variable and the end result rarely show a linear relationship. Kodama in early 1995 concluded that the deterministic model of the past for managing R&D processes does not apply anymore and it should be replaced with a model that blends the chaos theory and random factors. Thus, although it is simple and easy to exert control in accordance with the reliability requirement and control requirement for common processes, such an approach puts restraints on R&D units and curbs their ability to suggest creative solutions.

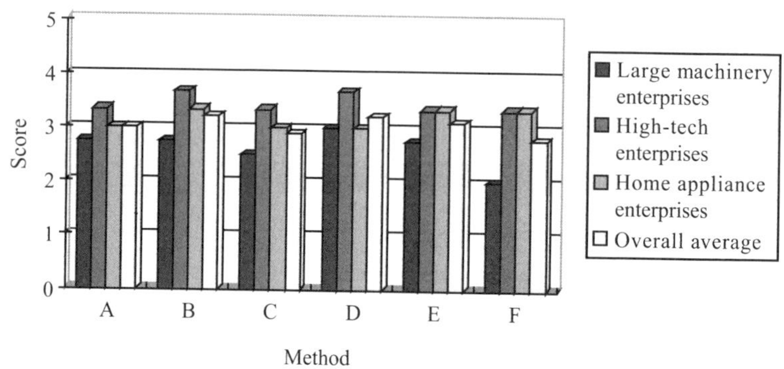

A: Annual R&D budget;

B: Various R&D units signing written assignment agreement;

C: Controlling in accordance with the reliability requirements or control requirements of a common process (e.g. a production process);

D: Developing a detailed plan as much as possible;

E: Standardizing research at various institutions as much as possible;

F: Controlling the entire R&D project through the overseas R&D unit which is in possession of key technologies and thus at the center of the R&D network.

Figure 4-11 Methods for controlling overseas R&D institutions

Considering the effectiveness of internationalized innovation process management in the surveyed enterprises (Figure 4-12), most successful projects were accomplished within the planned time frame, with a score of 3.6, but the R&D costs tend to exceed the budget, with a score of 2.6. Among them, for high-tech enterprises, projects can be completed basically within the planned time frame (scored at 4.33) and the situation is more satisfactory than that for large machinery enterprises (scored at 3.25) and that for home appliance enterprises (scored at 3.33). The number of cases where R&D costs are within the budget is the biggest for high-tech enterprises among the three industries with an average score of 3, as compared to 2.5 for large machinery enterprises and 2.67 for home appliance enterprises.

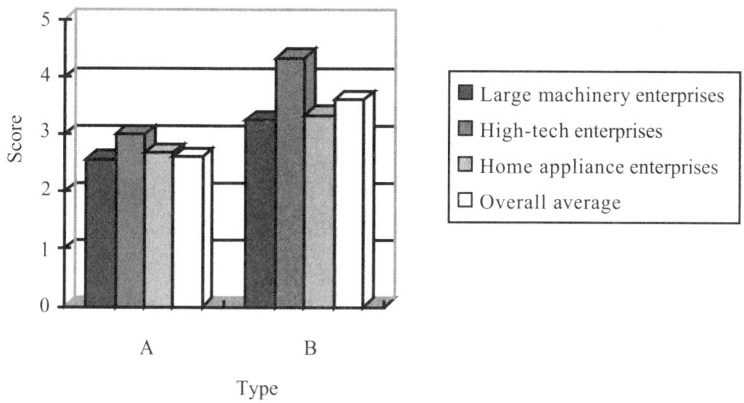

A: Innovation projects are completed within the budget;

B: Innovation projects are completed within the planned time frame.

Figure 4-12 Comparison of the actual process and planned process of innovation projects conducted by overseas R&D branches

4.3 Suggestions on internationalization of technological innovation for Chinese enterprises

By summing up previous studies by others and the analysis of our own, we found that there are two challenges to organizing and managing internationalization of technological innovation: first, corporate management must find an organizational model suitable for decentralized technological innovation activities; second, they must create the conditions for collaborative innovation carried out by multiple units and in multiple ways. Management by Chinese enterprises of their internationalized innovation processes still needs to be improved and perfected.

4.3.1 Evolution mechanism of the organizational form

Much literature on internationalization of innovation asserts that an innovation network enables an enterprise to acquire relevant knowledge from the market, new technology, and even its competitors; organizational structure is built on the principle that various units are interdependent; such interdependence between resources and responsibilities breaks the layer barrier that separates the local interests and global interests, and the approach is the self-implemented sharing of resources, ideas, and opportunities. Only when the units connect together based on the interdependent relationship rather than on the one-way dependent or independent relationship, can the unity be formed and operate an enterprise effectively.

Through statistical analysis, we found that there were no significant differences among the three industries in their evaluation of various organizational forms, but the scores received by the five organizational forms were quite different. After considering the present state of internationalization of technological innovation conducted by Chinese enterprises and the strategic development paths chosen, we believe that the organizational form for innovation under the condition of internationalization will also go through the following evolutionary paths.

• The initial stage of internationalization of technological innovation: home country-concentrated center-edge form

This model is mainly for the current situation where the general internationalization level of Chinese enterprises is not high, and these enterprises are still used to carrying out relatively centralized innovation. The main function of the R&D branches established abroad is technology monitoring. In other words, during the entire innovation process, the overseas R&D branches participate only during the innovation concept formation stage; as for activities such as project decision, product development, and problem-solving, they are all done at the domestic R&D center. To an innovation institution with such an organizational form, whether or not they can successfully carry out innovation activities abroad is not important, whereas what is important is to be able to make use of the existing technological achievements or the innovative ideas from abroad.

Such concentration of knowledge and technology under the center-edge form ensures that the company can quickly create new products and processes, but the R&D center in the headquarters lacks understanding of the needs of overseas markets and actual production there, and their feedback may not be able to meet the needs of all. Overseas branches with the role of technology monitoring points, due to the limitation on resources, have only a diminished number of opportunities for learning (Figure 4-13).

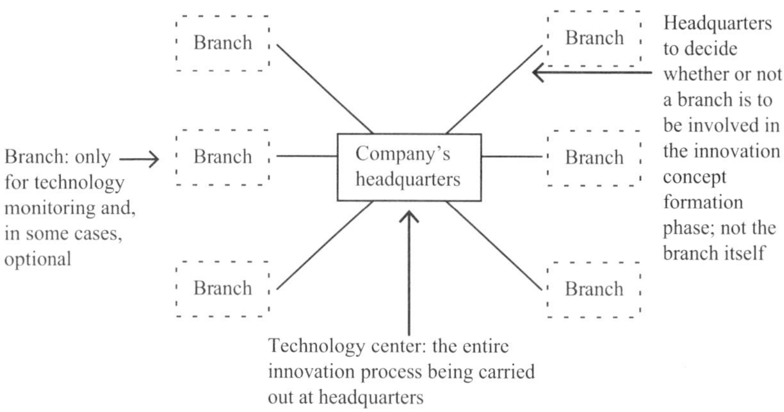

Figure 4-13 Home country-concentrated center-edge form

When using this organizational form, there must be emphasis on the control of overseas branches. As mentioned earlier, the surveyed enterprises gave similar scores to each of the methods of control listed in Figure 4-11. The statistical analysis shows that there are no significant differences among the enterprises with regard to the choice of methods for controlling internationalized innovation processes and there are no significant differences among the average scores of the various industries and the group as a whole. Through correlation analysis, we found the relationship between the use of center-edge organizational forms and the use of various control methods among Chinese enterprises (Table 4-2).

Table 4-2 Center-edge organizational forms and control methods

Control method	Correlation coefficient	Level of significance
Annual R&D budget	0.369	0.368
Various R&D units signing written assignment agreement	–0.007	0.987
Exerting control as for a common process (n)	—	—
Developing a detailed plan as much as possible	–0.046	0.913
Standardizing research at various institutions as much as possible (n)	—	—
R&D system with overseas R&D institution in possession of key technologies at the center position	–0.205	0.626

*Correlation is significant at 0.05 level (two-tailed).

Note: Projects marked "n" are those with unsatisfactory methods for innovation process control and will not be discussed here.

Under the center-edge model, no control methods show significant correlation, which is consistent with the actual situation. Obviously, choosing the method of control to use also depends on the effect of control, which will be further analyzed later.

With an ever-increasing technology level of Chinese enterprises, and as a growing number of Chinese enterprises start overseas expansion, such enterprises will certainly care more about whether or not their products can meet the needs of overseas market; thus, overseas branches will gradually start to undertake more product development tasks that are aimed at the local market and will provide technology support for further expansion into the local market by their enterprises. Thus, not long from now, the home country-concentrated center-edge model will no longer be the organizational form for internationalized innovation processes being carried out by Chinese enterprises.

• The development stage of internationalization of technological innovation: composite-star structure

Within the next few years, this model will become the main organizational form adopted by Chinese enterprises for carrying out internationalized innovation processes. Under this structure, the overseas R&D units will not only engage in technology search but also work on technology transfer to overseas countries or on product development for the market serviced by a subsidiary there; thus, the overseas branches will gain a higher status. However, the overseas R&D units will still rely on headquarters for resources and instructions, decision-making and control of innovation activities will still be centralized to a large extent, and communication among the overseas units will be minimal (especially face-to-face communication).

In addition, the choice of organizational structure will inevitably affect the effectiveness of internal communication. Therefore, now that horizontal communication is becoming ever more important in an international innovation project, the use of a purely center-edge organizational structure will inevitably lead to low efficiency for some dispersed global technological activities. To mitigate this problem, it is necessary to establish a parallel structure for the R&D units—a committee directing internationalized innovation activities—operating concurrently with the hierarchical structure (Figure 4-14).

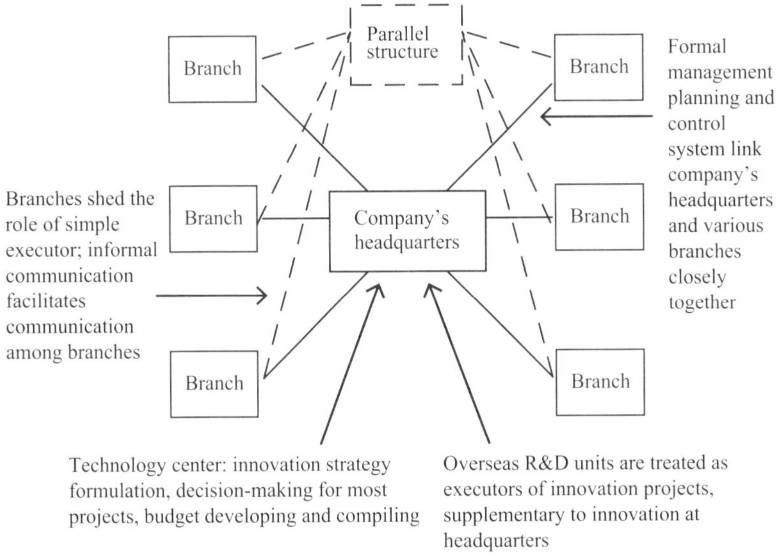

Figure 4-14 Composite-star structure

An informal work environment and a certain degree of autonomy are important for product development. Especially when the innovation activities are for overseas market, more autonomy should be given to local branches, so as to enable them to develop products/services suitable for the local market. A committee operating in parallel will not only play an important role in innovation activities but also provide a platform for information transfer inside the company. Members of the committee can be managers from overseas subsidiaries, R&D branches, and the company's central research institute. The main functions of the committee include: (i) to participate in the decision-making process of an innovation project, (ii) to participate in the planning of company innovation strategy, and (iii) to participate in supervising the work at various R&D branches. New product development can be carried out at the technology level and market level simultaneously. Since the committee does not have the power to make final decisions, the real value of such activities is in the formation of an informal network. The committee members will meet face-to-face regularly and thus promote communication among various branches. In the early period of innovation activities, informal and divergent methods are usually needed to obtain creative ideas. Overseas research indicates that ideas for new product development from employees, suppliers, and competitors are often easy to effectuate. Therefore, another advantage of using a parallel structure is that ideas from various functional departments can be consolidated during the initial period of a project, support can be secured from finance department and business department, and thus later activities will be better coordinated.

While encouraging the possible autonomy and enthusiasm of overseas branches, it should not be ignored that there must be a level of control over them. We believe that a company should only take the control measures that are most needed and not all possible measures, so as to achieve its objective with less cost. Through correlation analysis, we found the relationship between the choice of composite-star structure and the corresponding method of innovation process control (see Table 4-3).

Table 4-3 Composite-star structure and control method

Control method	Correlation coefficient	Level of significance
Annual R&D budget	0.094	0.398
Various R&D units signing written assignment agreement	–0.733*	0.008
Developing a detailed plan as much as possible	–0.713*	0.010
R&D system with overseas R&D institutions in possession of key technologies at the center position	–0.242	0.250

*Correlation is significant at 0.05 level (two-tailed).

Note: Two control methods are unsatisfactory and thus not listed here.

It can be seen that there are significant negative correlations between the methods of control by the use of written agreement as well as by laying down a detailed plan and the composite-star structure; thus, we believe that these two methods of control are not suitable for the composite-star organizational form and should not be used too much. Foreign research literature asserts that these two methods are helpful for cross-functional communication and for clearly defining the team responsibilities. Thus, to clarify responsibilities and avoid conflicts, each branch needs to sign some form of project agreement (Moenaert et al., 2000). However, this is the approach to adopt for strengthening control when an organization attaches importance to horizontal communication. In the next few years, the Chinese enterprises will use composite-star structure for internationalized innovation activities more often and top-down communication and control will be the primary approach. In many cases, responsibilities and plans for overseas R&D units will be directly issued by the company's headquarters. These two methods of control will increase the cost for coordination among the various units, causing unnecessary confusion.

Although control by the use of annual R&D budgeting and through the overseas technology centers of excellence in possession of key technologies do not show strong correlation with composite-star structure either, we still believe that these two can be used in Chinese enterprises with star organizational structure compared with the other methods, especially when controlling through the establishment of overseas technology centers of excellence. On the one hand, the scope and cost of control by the headquarters can be reduced and autonomy of overseas R&D institutions can be increased; on the other hand, such a structure can lay a foundation for later organizational transformation into a globally interconnected network model.

· The mature stage of internationalization of technological innovation: globally interconnected integrated innovation network

Overseas research literature has shown that globally interconnected integrated innovation network has become a trend of an enterprise's innovation organizational development (Von Zedtwitz, 2002). Internationalization of technological innovation highly requires mobile coordination, and integrated resource network is of vital importance to flexible and complex responsiveness for cross-border innovation development. Technological innovation organizations have become more open, differentiation among various organizational units and specialized capabilities have made it certain for such units to cooperate in the creation of new products and new processes, and interdependence among the various units makes such cooperation self-sustainable.

Obviously, building such an innovation network often requires the company to have great strength; at the same time, to the company, the international market has become as important as domestic market and even more important. Thus, at this point, the effort of enterprise technological innovation should be directed specifically to the international market. In order to achieve global competitive advantages, enterprises should concentrate on some resources in the home country, concentrate on some resources abroad, and even scatter some resources for the operations in various countries; as a result, a complex structure where assets and capabilities have all gone through segmentation and specialization will be formed. In addition, dispersed resources will be integrated through interconnection inside the company (see Figure 4-15). Some interdependencies are also the natural results of such a structure, where assets and capabilities are segmented and divided among the various units. Such relationships are usually intentionally designed so as to ensure that connected branch units will consciously cooperate with each other.

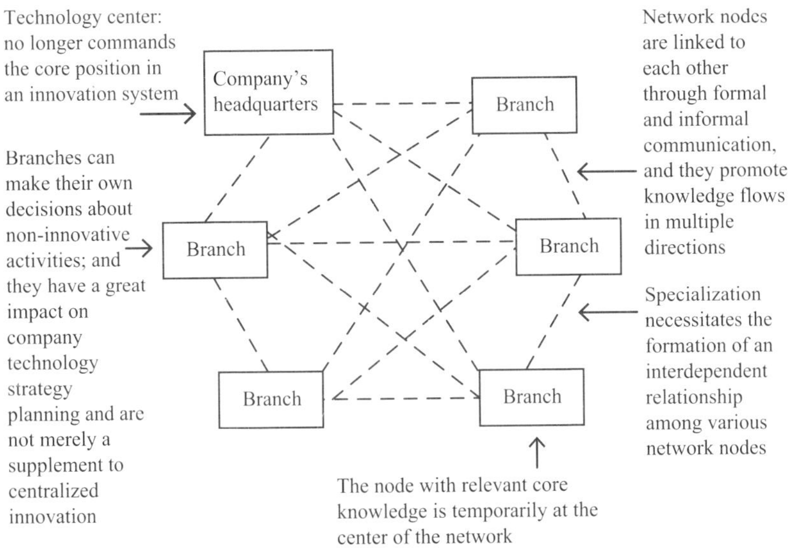

Figure 4-15 Globally interconnected integrated innovation network

In this mode, technological innovation may need to be carried out in a cooperative manner, by the R&D units around the globe; domestic R&D center or central research institute is no longer at the core position, and an absolute core node does not even exist. For each different innovation project, an R&D unit with relevant core knowledge will be at the center position of the network and will maintain control of the project partners. Branches can also decide on their own core knowledge to carry out R&D projects and seek financial support for such projects, but decisions for innovative projects of strategic importance to the company need to be made jointly by multiple R&D units. In addition, the company's headquarters exercises strategic control at the macro level over technological innovation activities of the whole company.

However, while network structure can promote the free flow of products, resources, information, and manpower, it is very difficult to set up effectively due to the various formal and informal systems in existence and it requires very high management costs as well as advanced management ideas, excellent managerial skills, and resolute decision style. Even among world-class large multinationals, few can successfully put this model into operation.

As shown in Table 4-4, Chinese enterprises also do not have a clear understanding about control within an integrated innovation network. Integrated innovation networks—for Chinese enterprises that are relatively backward in terms of technology and economic strength—can only be at the stage of theoretical discussion. We believe that as internal control exerted by the company's headquarters is greatly weakened, the smooth transitions between the successive stages of innovation activities under many circumstances need to be coordinated by the various branches themselves; thus, approaches such as clearly defining the responsibilities of each branch (signing written agreements) should become the control method under the network mode.

Table 4-4 Integrated innovation network and control method

Control method	Correlation coefficient	Level of significance
Annual R&D budget	0.280	0.501
Various R&D units signing written assignment agreement	–0.137	0.746
Developing a detailed plan as much as possible	–0.344	0.404
R&D system with overseas R&D institutions in possession of key technologies at the center position	–0.494	0.214

*Correlation is significant at 0.05 level (two-tailed).

Note: Two control methods are unsatisfactory and thus not listed here.

4.3.2 Exchange and communication

There is no doubt about the importance of exchange and communication for internationalized innovation activities, and thus, enterprises should do their best to promote effective communication in the innovation process. In addition to creating the necessary conditions for effective communication in the design of organizational structure, the following should also be done to enhance the enterprise's capabilities.

- **Communication infrastructure**

In recent years, the rapidly developing information resources and information technologies have provided new opportunities for managing cross-border innovation. Videoconferencing systems and remote log-in systems provide relatively effective communication at very low additional costs, make possible a way of information sharing that did not exist previously, and thus make available more opportunities for information exchange among the team members. It can be said that without modern information and communication technologies, there would not have been cross-border innovation projects. Therefore, in order to achieve bona fide internationalization of technological innovation, Chinese enterprises must make use of modern communication technologies to build powerful internal information networks and accomplish easy flow and sharing of knowledge.

- **Development of trust**

Being consistent with the research reported in overseas literature, our case studies and questionnaire surveys also support this conclusion: information technologies can facilitate the exchange of explicit knowledge, and they have gradually taken the place of people traveling back and forth among various branches for communication purposes, but still the relations and trust among employees form the foundation of effective communication. Knowledge embedded in technological innovation processes and experiences learned by the project team members are more important than written and explicit knowledge, and tacit knowledge can only be communicated face-to-face. In addition, certain level of frequency for exchange should be maintained; otherwise, the degree of trust between two parties will decrease and communication will become meaningless. Today's electronic communications are very efficient in processing routine matters, but they are powerless in handling interpersonal and emotional problems and can easily lead to leakage of secret. Without face-to-face communication, it would be more difficult to eliminate cultural differences, especially at the early stage of a project when the team spirit is still in the making and trust among team members is very important. Trust should be developed from the following aspects.

(1) Participation is a crucial element of confidence building. Communication determines the attitude and behavior of each participating team. Decisions made by those who are widely supported in the network can be better implemented, and without extensive participation, it will be difficult for various parties to achieve agreement and harmony. Participation in formal meetings and decision-making processes is only the first step in creating an atmosphere of trust. The top management should also make employee participation in the management of international innovation projects a part of the corporate culture. In full-length participation for managing such projects, person-to-person cooperative relationship will be formed and confidence with higher-level management will be established.

(2) Creating organizational openness will contribute to the assurance of a fair and impartial system within the organization. Fairness cannot be achieved simply by designing a good organizational structure; it must be strengthened with the discipline of language and behavior, especially for leaders of the various teams. Decisions about increase or decrease in the number of overseas branches, innovation projects, assignments for each branch, and so on must be made in a transparent and rational communication network and should not be done by "black box operations." Such decision-making may bring about some conflicts, but the feeling of fairness as a result of such process can protect and enhance the employees' confidence in the company.

By analyzing the correlation between organizational openness and communication effectiveness as well as factors of innovation project failure, we can attain further understanding about the benefits of organizational openness (see Table 4-5).

Table 4-5 Effect of different approaches in creating organizational openness

			A	B	C	D	E	F	G
Communication effectiveness	Between the parent company and the branches	Correlation coefficient	0.430	0.387	0.635*	0.462	0.733*	0.212	0.374
		Significance	0.107	0.135	0.024	0.089	0.008	0.278	0.144
	Among branches	Correlation coefficient	0.685*	0.424	0.690*	0.461	0.560	0.780	0.542
		Significance	0.021	0.128	0.020	0.106	0.058	0.007	0.066
Communication factors in innovation project failure	Communication difficulties between foreign researchers and Chinese researchers	Correlation coefficient	–0.535*	–0.573*	–0.506	–0.193	–0.452	–0.392	–0.071
		Significance	0.045	0.033	0.056	0.284	0.081	0.116	0.418
	Communication difficulties between the parent company and the branches	Correlation coefficient	–0.418	–0.648*	-0.423	–0.152	–0.665*	–0.183	0.055
		Significance	0.100	0.016	0.098	0.328	0.026	0.295	0.436

*Correlation is significant at 0.05 level (two-tailed).

Note: For specific meanings of A–G, see Figure 4-5.

Thus, we conclude that Chinese enterprises so far have adopted different approaches to create organizational openness and such different approaches have produced different results. In the next few years, in the process of internationalization of technological innovation, enterprises should strive to promote regular contact between managers and R&D personnel in overseas R&D branches (Approach A), to promote exchange and communication between managers as well as managerial staff at overseas R&D branches and people at R&D units of the parent company's headquarters (Approach B), to hold formal and informal seminars regularly at overseas R&D branches (Approach C), and to organize international teams as well as encourage researchers from different countries to establish personal contacts and influence each other (Approach E). Obviously, many domestic enterprises have already begun to put these into practice. Presently, the other three approaches, such as D, F, and G, used most often by Chinese enterprises do not appear to have produced the claimed results; thus, we suggest that they should be used only when an enterprise is rich in the needed resources and has plenty of manpower and material capabilities. It should be noted that foreign literature often proposes to effectuate tacit knowledge exchange by job rotation, moving R&D personnel around, and so on, but Chinese enterprises have obviously not paid much attention to such approaches. Probably because these Chinese enterprises have just started to internationalize their technological innovation, they have not thus far recognized the importance of such approaches. However, in order to develop an integrated innovation network, it is a must to recognize that the exchange of tacit knowledge is a very important factor.

(3) Project manager's ability facilitates the confidence building process. No matter how technology develops and no matter what new management approach emerges, people will always be at the center of innovation. The rapid increase in the amount of technological information and the ever-increasing trend of globalization require us to give more attention to managing knowledge and human resources.

International innovation projects require professional project managers with excellent skills. Such managers are able to adapt to different cultures, meet the needs of different people, and get the support of team members. In short, project managers are the driving force of an innovation process. If an inexperienced project leader is chosen, he/she may not merely spoil cooperation for the project but oftentimes even lead the project into difficulty from the very start. At this stage, many Chinese enterprises choose project managers by the standard, namely, "the most easily found, not the most suitable"; technological expertise alone is far from enough for leading an innovation project. For enterprises that are preparing for internationalization of technological innovation, the first task of human resource management is choosing the project manager.

· Balancing communication by IT tools and face-to-face communication

As analyzed previously, communication by IT tools is not yet a perfect substitute for face-to-face communication and learning tacit knowledge still has to be done through direct communication. However, to use face-to-face communication as the main form of communication, costs (including time cost and financial cost) are the major obstacle to most enterprises. Thus, it is necessary to find a balanced solution to achieve effective communication.

In the early stage of internationalization of technological innovation, the enterprise is oriented toward the domestic market, almost all innovation activities are carried out domestically, and knowledge flow between overseas branches and company's headquarters is oftentimes one way and simple, that is, the overseas branches follow the instructions from the company's headquarters and transmit technology trend information gathered by them—especially trend information about the knowledge or technology adopted by the overseas enterprises that are competing in the domestic market with the enterprise—to the technology center of the enterprise, and then domestic R&D center will set up R&D projects accordingly.

Meanwhile, it is rare for such knowledge to flow among the overseas branches. Obviously, much of the knowledge is explicit knowledge that can be transmitted in a standardized form. Constructing information infrastructure and network and maintaining channels of contact through information technologies are the focus during this stage. Face-to-face communication—when the content of communication is mainly of explicit knowledge—is not really more effective than communication through IT tools. Meanwhile, due to the considerations for cost saving, except in cases such as routine work reporting or special circumstances, face-to-face communication does not need to be given special attention during this stage.

In the development stage of internationalization of technological innovation, overseas branches have begun to undertake some product development tasks and the knowledge that they create—under some circumstances—will be transferred to the domestic technology center, which will greatly increase the complexity of communication. At the same time, although the knowledge flow at this stage is mainly two-way between the company's headquarters and branches, knowledge transfer among the branches themselves have begun to appear. To ensure the possibility of organizational learning, ABB company's experience is a lesson worth looking at. ABB adopts a top-down technology strategy; meanwhile, it sets up a technology management team (TMT), which develops operational plans for the company's R&D, and clarifies responsibilities of the various business departments. The company's low-level teams and TMT communicate face-to-face, but coordination within each team and coordination within TMT are done via IT tools. The direct contact between the different levels is maintained by a person especially appointed for this task. This model brought huge success for ABB. In the results of our questionnaire survey, there are no significant differences among the three industries in the scores they gave to each of the three items, "communication between overseas branches and company's headquarters mainly done face-to-face," "communication between overseas branches and company's headquarters mainly done via IT tools," and "communication among overseas branches mainly done via IT tools". However, high-tech enterprises attach more significance to face-to-face communication among overseas branches than the other two industries. For communication during the cross-border innovation process, Chinese enterprises may adopt the ABB model. The high-tech enterprises can give their parallel committees greater power so as to facilitate face-to-face communication among the overseas branches.

In the mature stage of internationalization of technological innovation, on the one hand, because each overseas branch is highly specialized and there is no longer an absolute center unit in the organization, an innovation project usually requires close cooperation among various branches and thus exchange of experience and communication about work progress among various branches become very important. With the emergence of more advanced IT tools (even the sense of touch may be transmitted) and the changes in social attitudes, the differences in the effectiveness of communication via IT tools and face-to-face communication will further decrease and the broader use of IT tools will still be the main trend in communication. On the other hand, in this stage, the flow of knowledge is multidimensional, knowledge transfer among various network nodes will become the main communication, exchange of tacit knowledge will account for a greater share in the total content transfer and will often be crucial to the smooth advancement of the project, and thus the benefits of face-to-face communication will still be far greater than those from communication via IT tools. After entering the mature stage of internationalization, strength of the enterprise has very much increased and has the financial resources for more extensive face-to-face communication. Therefore, at this stage, enterprises should emphasize face-to-face communication and IT-enabled communication equally.

4.3.3 Integration of internationalized innovation processes

- **Project planning phase**

The main activities of this phase are to propose innovative concepts, develop detailed solutions, and make the final decision on the choice of solution.

The sources for innovative ideas are diversified, and enterprises must integrate the creative ideas from each of them: customers, employees, competitors, distributors, general distributors, suppliers, and so on. Overseas branches enjoy proximity to the local markets, which is different from the case where innovation is concentrated in the home country, and thus their ideas are much closer to the needs of the local markets. The market needs of various regions worldwide are often very different from each other. Meanwhile, with an increase in the number of overseas branches and subsidiaries, more and more new ideas will be suggested. Developing customized products for each specific market segment or even each specific client can satisfy their special needs, but this approach will increase the costs, where the enterprises will be in a dilemma. Standardization is easy for management purposes, but the products may not meet the needs of the local markets. Therefore, for the Chinese enterprises with not-so-advanced technologies, the following principle is necessary: core components that are not usually seen should be developed at the company's headquarters and produced with standardized production, but those components that are seen should be modified for the local markets by the local branches.

After an innovation concept has been proposed, its details need to be worked out and final choices need to be made. During different stages of an international innovation project, overseas branches will have different decision power; however, whatever the decision model is, a company's innovation activities must be a diversified process with feasible and effective participation from various units and groups;

there should not be any bias toward any particular business, function, or region; and the decision-making process of the company should give every idea an opportunity to express even though each will have a different level of impact. Thus, it is of particular importance to involve non-R&D departments in the project as early as possible. The questionnaire survey shows that Chinese enterprises do not do well in this regard and they should seek to change the situation as soon as possible.

- **Project implementation phase**

The main activities of this phase are to set up the innovation team, promote smooth cooperation among various branches, and closely watch over the progress of the project.

Goal congruence of the various R&D units is one of the major factors that will bring about team cohesion. The use of performance indicators for rating local business units impedes the formation of team cohesion. Because consistent long-term goals are lacking, local units will focus on local business operations and short-term projects so as to meet the evaluation standards. Moreover, because of these evaluation standards, marketing department and sales department will not be willing to take risks either. Product and marketing managers of the multinationals change very frequently; thus, they usually refuse to undertake radical innovation projects. Every unit should formulate specific goals and strive to achieve optimization for the company as a whole and not for the local unit itself.

Now, the cost of new technology development is getting higher and higher. As a result, many multinationals no longer emphasize the implementation of deep R&D internally for technological innovation and, instead, they turn to the outside to seek technological innovation capabilities (Von Zedtwitz, 2002). Another result of the high cost is that when many technologies show great potential, the company can choose only one to concentrate its limited resources on. Many enterprises have discovered that it is very difficult to develop several different technologies simultaneously; thus, in today's environment when technologies rapidly develop, cooperation with outside laboratories is crucial for an enterprise's success. Many Chinese enterprises only regard cooperation as a tool for entering into a market or acquiring certain technologies. Moreover, we have found through the survey that many enterprises still consider outsourcing some business functions as an operational problem—the standard for selection is still the lowest bid and the outsourcing decision is not based on strategic considerations. In short, although cooperation with external forces is of strategic importance, such cooperation has so far rarely been included in the horizon of strategic planning. Obviously, this is not unrelated to the relatively low level of technology used in the Chinese enterprises.

In technological innovation, an enterprise will not only want a project to achieve its goals but also want to control the cost and time needed within the planned range. Considering the situation of the surveyed enterprises, the relationship between the control method and the result can be seen in Table 4-6.

Table 4-6 Control method and the result

Control method	Result			
	Project completed within planned costs		Project completed within planned time frame	
	Correlation coefficient	Level of significance	Correlation coefficient	Level of significance
Annual R&D budget	0.559*	0.047	0.287	0.227
Various R&D units signing written assignment agreement	0.409	0.137	−0.065	0.429
Developing a detailed plan as much as possible	0.509	0.067	0.007	0.492
R&D system with overseas R&D institutions in possession of key technologies at the center position	0.723*	0.009	0.227	0.264

*Correlation is significant at 0.05 level (two-tailed).

Note: Two control methods are unsatisfactory and thus not listed here.

With an analysis of considered organizational forms, we conclude that during different phases of internationalization of technological innovation, an enterprise should adopt different control methods. In the initial stage of innovation internationalization, the number of overseas branches is very small, the functions entrusted to them are few, and such branches cannot form overseas technology centers of excellence; their main financial support comes from the headquarters. Thus, annual R&D budgeting should be the main control method during this stage. By the development stage, as analyzed earlier, composite-star structure will become the main organizational model and enterprises using this organizational model are not willing to use written agreements and develop detailed plans as control methods. Table 4-6 once again emphasizes that setting up overseas technology centers of excellence and annual R&D budgeting should be the control methods for Chinese enterprises. Therefore, we believe that enterprises should set up overseas technology centers of excellence for controlling overseas R&D units and should use annual budgeting as a supplement to that approach. After entering into the mature stage, the control of the branches by the company's headquarters will be greatly weakened, methods to ensure that an innovation process proceeds smoothly will not be limited to the choices we provide, and thus more control methods other than signing written agreements will appear.

In addition, even though the relationships between factors, such as dividing a large project into subprojects, dividing project implementation into phases, and maintaining synchronization among the various R&D units in work progress, and the innovation processes are not reflected in the questionnaires we collected, research of overseas literature has pointed out that when internationalization of technological innovation reaches a certain stage, enterprises will be confronted with such issues; thus, Chinese enterprises should start to pay attention to such questions now and put the solutions into practice according to the actual situation.

- **Market introduction phase**

The main activity of this phase is to move a technologically accomplished project to the market and strive for consumer acceptance of its products.

Choices made during the market introduction phase will to a large extent determine the outcome of the project—success or failure. Many enterprises are aware of the potential benefits of being a pioneer in the market, but they also know the costs and risks involved with such a strategy. The potential benefits include less difficulty in setting up distribution channels, more shelf space, natural building up of barriers for latecomers, and so on. In addition, customers can readily accept the early brands. However, revenue is always proportional to the risk and cost involved, and both new product development and marketing are very expensive; once a new product has been introduced into the market, the impact of project failure will even exceed that of a technological failure at the development phase. Obviously, there are also many enterprises that choose to be latecomers. One potential advantage of being a latecomer is that the demand for the new product has already been activated; obviously, in order to exploit this potential advantage, the latecomer enterprise must possess a certain level of strength: better technology, better brand recognition, adequate production capacity, and extensive distribution channels.

In our survey, there are no significant differences among Chinese enterprises in the choices they make during the market introduction phase. However, based on the research literature and case studies, we have come to the belief that with the present strength of Chinese enterprises, in the initial phase or development phase of internationalization of technological innovation, it is better to enter a market after demand for a product has already been activated, when the market begins to enter the growth period, and the barrier of entry is not yet too high. In a mature market, the competition is usually intense and, as a result, product development costs cannot be quickly recovered.

However, the risk and initial cost for being a market pioneer are pretty high, and such situations may not necessarily bring about good results for Chinese enterprises, which are not that powerful. Meanwhile, the development experiences of several Chinese well-known home appliance enterprises have also proven to us that entering a market immediately after the market pioneer can often bring greater benefit to enterprises with an average strength. Once an enterprise enters into the mature stage of internationalization of technological innovation, its strength will gradually grow and then it can make strategic adjustments accordingly and freely choose whether to be the market pioneer or a follower in introducing a new product.

For the model that is based on centralized technological innovation and production, product export is the primary method for market introduction when an enterprise is at the initial stage of internationalization of technological innovation; this is also the primary means for Chinese enterprises to push their products into overseas markets. Although this approach obviously brings the benefit of the economy of scale, it cannot respond to the timely needs of the local market and there are extra expenses for product transportation. In the future, when the innovation model becomes geographically dispersed, there will be a need to be more advanced in the approaches to introducing new products in the market.

After correlation analysis (see Table 4-7), we conclude that when entering a market after the market has been initially activated, wholly owned subsidiary approach should be adopted to do it and other methods can be used for supplementary choices. With this approach, large-scale production can be carried out locally to meet the ever-growing market demand—that has been activated in that region—for the new product. Meanwhile, the control requirement of the company will also be met. Thus, in the development stage of internationalization of technological innovation, the wholly owned subsidiary approach should be the primary means for introducing innovative products into the local market. In the mature stage of internationalization of technological innovation, overseas technology centers will take more responsibilities; fruits of innovation are no longer limited to new products or new processes, including patents, documents, and so on; and as a result, the methods for commercialization will naturally become diversified.

Table 4-7 Timing and methods of market introduction

		Wholly owned company	Joint venture	OEM	Product export	M&A	Patent transfer
Developing new market	Correlation coefficient	0.340	0.370	–0.244	–0.153	0.283	–0.366
	Significance	0.336	0.293	0.498	0.694	0.452	0.372
Entry after the market has been initially activated	Correlation coefficient	0.667*	0.571	–0.100	0.437	0.365	0.112
	Significance	0.035	0.085	0.783	0.240	0.335	0.792
Entry when the market has matured	Correlation coefficient	0.461	0.136	–0.064	0.585	0.568	–0.080
	Significance	0.180	0.708	0.860	0.138	0.111	0.850

*Correlation is significant at 0.05 level (two-tailed).

4.4 Chapter summary

With the development of economic globalization and China's accession to the WTO, technological activities of Chinese enterprises will also integrate with the global technological trends. How to manage innovative activities during different stages of internationalization of technological innovation has become an issue that Chinese enterprises urgently need to know about (Table 4-8).

The characteristics and boundaries of organizations play a very important role in the process of technological innovation. Integrated innovation network is the trend of enterprise innovation activities; however, at this stage, Chinese enterprises still emphasize centralization and control and, thus, the star structure will be the main structure of cross-border innovation in the near future.

Managing knowledge and directing information flow

inside an organization are all done with the innovation process as the center; thus, it is necessary to enhance communication and connections among innovation branches, between the company's headquarters and branches, and between R&D units and non-R&D units. Constructing communication infrastructure, developing trust, and attaining balance between communication via IT tools and face-to-face communication are the problems to be dealt with in cross-border communication.

In the internationalized innovation process, choosing the right innovation projects, ensuring full cooperation of each participating department, and finding the opportunities and methods for commercialization are all worthy of special attention by the companies.

Table 4-8 Internationalization of technological innovation and organizational management

Features of organizational management	Stages		
	Initial stage of internationalization of technological innovation	**Development stage of internationalization of technological innovation**	**Mature stage of internationalization of technological innovation**
Organizational form	Home country-concentrated center-edge form	Composite-star structure	Globally interconnected integrated innovation network
Innovation activities of overseas branches	Formation of innovative concept	Limited product development	Major technological innovation
Centralization, decentralization and decision-making	Absolute centralization	Participatory centralization	Decentralization
Knowledge development and diffusion	Development and possession by headquarters	Development by headquarters and transfer to overseas branches	Decentralized development and global sharing
Exchange and communication	Building infrastructure for communication network and developing trust for it		
Knowledge flow	Unidirectional and simple	Bidirectional and complex	Multidirectional and complex

(Continued)

Features of organizational management	Stages		
	Initial stage of internationalization of technological innovation	**Development stage of internationalization of technological innovation**	**Mature stage of internationalization of technological innovation**
Balance between IT-enabled communication and face-to-face communication	IT-enabled communication in the main, with regular face-to-face communication	IT-enabled communication among overseas branches, and face-to-face communication between the company's headquarters and overseas branches	Comprehensive and extensive promotion of IT-enabled communication and face-to-face communication
Process integration	Integrating creative ideas from all parties, building appropriate project teams, and promoting cross-functional cooperation		
Control method	Annual R&D budget	Establishing overseas technology centers of excellence and annual R&D budgeting	Signing written agreements
Market introduction	Entry after the market has been initially activated and product export	Entry after the market has been initially activated, setting up wholly owned subsidiaries	Flexible timing for market introduction and variety in approach

Many companies regard the use of innovation as a core strategic capability. The competition has shifted to whether or not a company can keenly discover new trends, respond to them creatively, and spread its innovations worldwide, that is, the company's ability to engage in global learning. With the ever-increasing internationalization of technological innovation, decentralized innovation activities have posed challenges to the traditional organizational management.

After a long period of development, most companies have formed their own tradition in organizational management. For example, the early success of Panasonic is attributed to the cost advantage brought about by the scale of its centralized R&D and production in Japan, while Procter & Gamble gained success through the use of its powerful technologies and special marketing prowess. As a result, these companies will each lean toward the traditional specialty of its organization. Interdependency—which comes with developing multidimensional capabilities and geographically dispersed organizational forms—has transformed the system of control for companies originally with one-dimensional capabilities. With the ever-increasing internal differentiation and specialization of organizational roles as well as management methods, it is no longer enough for organizations to limit their approaches to reallocate resources and fine-tune control methods; they now must develop new comprehensive integration mechanisms encompassing areas from organizational model to management method. For internationalized innovation activities, geographic distance will give rise to tremendous centrifugal pull in an organization and the full play of creativity requires that top-down control must be weakened; therefore, each company must integrate ideas and capabilities from both inside and outside of its organization, ensure smooth information flow through its internal communication channels, adequately consolidate professional capabilities of its various branches, increase their interdependencies, encourage knowledge sharing as well as various other inputs, and establish a mechanism of coordination for effective decentralized innovation activities.

References

Arimura, S. Global R&D management: the case study of Mitsushita Electrical Co., Ltd. and Sony Corporation, CD-Rom. Portland, OR: PICMET, 1997.

Bartlett, C. A. & Ghoshal, S. Managing innovation in the transnational corporation. In: Bartlett, C. A., Doz, Y. & Hedlund, G. (eds.). *Managing the Global Firm*. London: Routledge, 1990.

Behrman, J. N. & Fischer, W. A. *Overseas Activities of Transnational Companies*. Cambridge, MA: Oelgeschlager, Gunn & Hain Inc., 1980.

Brockhoff, K. *Internationalization of Research and Development*. Berlin: Springer-Verlag, 1998.

Gassmann, O. & Von Zedtwitz, M. New concepts and trends in international R&D organization. *Research Policy*, 1999, 28(2-3): 231-250.

Gerybadze, A. & Reger, G. Globalization of R&D: recent changes in the management of innovation in transnational corporations. *Research Policy*, 1999, 28(2-3): 251-274.

Golder, P. Insights from senior executives about innovation in international market. *Journal of Product Innovation Management*, 2000, 17(5): 326-340.

Leonard-Barton, D. *Wellsprings of Knowledge: Building and Sustaining the Sources on Innovation*. Boston: Harvard Business School Press, 1995.

Moenaert, R. K., Caeldries, F., Lievens, A. & Wauters, E. Communication flows in international product innovation teams. *Journal of Product Innovation Management*, 2000, 17(5): 360-377.

Nonaka, I. & Takeuchi, H. *The Knowledge Creating Company: How Japanese Companies Create the Dynamics of Innovation*. New York: Oxford University Press, 1995.

Von Zedtwitz, M. Organizing global R&D: challenges, dilemmas and future trends. Miami: IAMOT2002, The 11th International Conference on Management of Technology, 2002.

Chapter 5

Internationalization of Technological Innovation: Resource Allocation and Integration

In resource theory, an enterprise is a community made up of special resources and resources are to be transferred and created in this community. An enterprise's system for internationalization of technological innovation comprises R&D units dispersed in a number of geographic regions, and it is a multinational network within which resource transfer and creation are carried out. Resources are scarce and are not inexhaustible. Under the condition of globalization, enterprises do not compete on scale, but on the ability to globally configure resources and coordinate international activities. To effectively carry out international technological innovation, an enterprise not only needs to match its strategy with its organizational form but also needs to configure its resources and fit its resource allocation to its strategic organizational form.

There are three aspects, that is resource allocation, integration, and advantage creation, to be considered in the process of internationalization of technological innovation, and there are three challenges posed to enterprises by globalization.

(1) Effectively acquiring necessary resources globally. Technological innovation is the "new combination of production factors" and the combination of innovative resources (information, ideas, materials, personnel, etc.). Technological innovation at any level and of any size requires resource input. An important success factor for a multinational is its ability to identify the favorable resources in a particular country or region around the globe and to obtain the rights to use such resources. At the company level, internationalized technological activities optimally use the special resources of each of the different regions, interact with enterprises in the local areas, effectively configure business capabilities, and promote the creation and diffusion of innovation (Cantwell & Kosmopoulou, 2001).

(2) Effectively configuring resources around the globe. The level of efficiency in resource allocation limits the output rate of R&D

activities by multinationals. Under the condition of globalization, the company has to create a corporate culture and a value system, which allow it to move its resources to any place in the world in order to maximize its competitive advantages (Rhinesmith, 1996). Efficient resource allocation can effectively make use of limited resources, reduce the input of innovative activities, and improve the output rate of technological innovation.

(3) Effectively integrating resources around the globe. For a multinational to transform its resource advantages into competitive advantages, the key factor is to effectively integrate the special resources of the parent company with the special resources of the host country of each of its subsidiaries. The nature of global competition is the global integration of value-added activities (Ghemawat & Spence, 1986). Integration is a key dimension of corporate strategy, and the successful integration by a multinational of its decentralized operational processes has a significant impact on its overall performance (Flaherthy, 1986).

5.1 Capital investment and allocation for internationalization of technological innovation

With the emergence of the global knowledge economy, multinationals are no longer satisfied with the business model of past decades, where new products and processes are developed domestically and then the fruits of R&D get transferred to overseas subsidiaries and are adjusted to meet the needs of foreign markets. In recent years, increasing investment in overseas technological innovation has become an important component in the implementation of globalized business strategies by multinationals of developed countries.

5.1.1 Capital investment and allocation for internationalization of technological innovation

(1) Increase in capital inflow/outflow of internationalization of technological innovation

Multinationals, in order to implement their global business strategies, continue to increase spending on overseas technological innovation. For a specific country, the pace of development in spending for internationalization of technological innovation is characterized by two features: first, the increased spending by domestic multinationals on overseas technological innovation has been accelerating; second, the increased spending by foreign multinationals on the country's technological innovation has also been accelerating.

The United States is the main country for implementing global technological innovation, where the spending for it clearly demonstrates the abovementioned two features. According to the statistics of the U.S. Department of Commerce, from 1986 to 1997, the overseas R&D spending by U.S. multinationals has grown from $4.6 billion to $14.7 billion (Figure 5-1), which is twice as much as the increase in R&D spending by foreign multinationals in the United States during the same period. By 1997, overseas R&D spending by U.S. companies has increased from 6.4% of their overall R&D spending in 1985 and 9.7% in 1995 to 10.5%. During the same period, the growth rate of R&D spending in the United States by foreign companies and their subsidiaries was very high (Figure 5-2). According to a survey conducted by the U.S. Bureau of Economic Analysis under the U.S. Department of Commerce, from 1987 to 1997, the R&D spending by U.S. branches of foreign companies increased from $6.5 billion to nearly $19.7 billion—accounting for nearly 15% of overall R&D spending of those companies—and this increase exceeded the overall domestic R&D spending by U.S. companies. However, after 1996, the increase in domestic R&D spending by U.S. companies started to exceed the increase in R&D spending in the United States by non-U.S. companies (Dalton & Serapio, 1999).

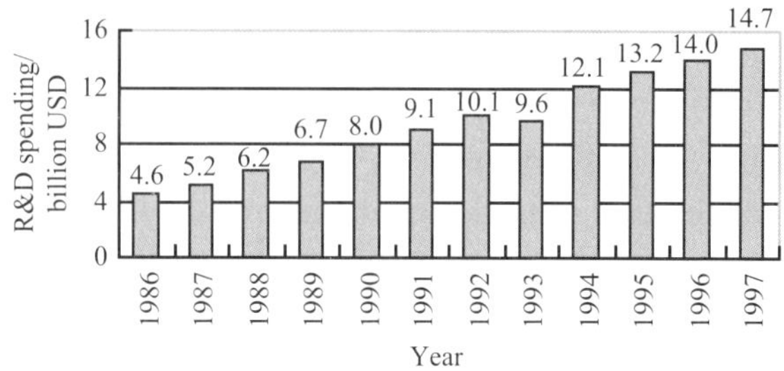

Figure 5-1 Overseas R&D spending by the U.S. multinationals

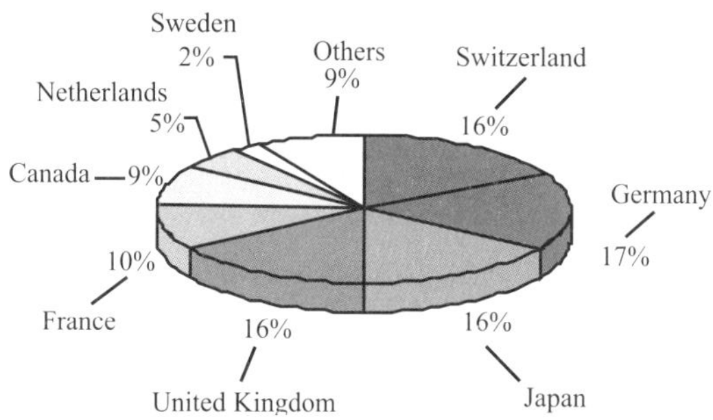

Figure 5-2 R&D Spending in the United States by foreign companies (Year: 1997)

Source: Dalton, D. H. & Serapio, M. G. *Globalizing Industrial Research and Development*. Washington, DC: U.S. Department of Commerce, Technology Administration Office of Technology Policy, 1999.

(2) Imbalance in capital inflow/outflow of internationalization of technological innovation

Internationalization of technological innovation is presently still in the initial stage of development, and it happens mostly among the "Big Triangle"—the United States, Japan, and the European Union. Thus, Patel and Pavitt (1998) believed that the internationalization of technological innovation should presently be more accurately described as "triadization" and not "globalization": the level of technological innovation and R&D activities is significantly higher than those in other countries; R&D capital inflow and outflow are both concentrated in these countries—displaying a pattern of cross-investment among the developed countries; and the R&D capital inflow and outflow among different countries differ greatly.

On the one hand, there is a big imbalance when multinationals choose the host countries for investment in technological innovation and R&D activities. In the example of overseas R&D spending by the U.S. companies, with the exception of Brazil and Australia, overseas R&D activities by the U.S. multinationals are mostly concentrated in the "Big Triangle" (Figure 5-3).

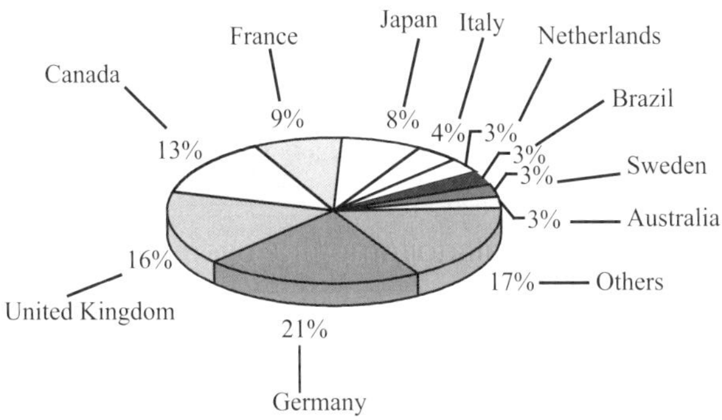

Figure 5-3 Structure of overseas R&D spending by U.S. companies (Year: 1997)

Source: Dalton, D. H. & Serapio, M. G. *Globalizing Industrial Research and Development.* Washington, DC: U.S. Department of Commerce, Technology Administration Office of Technology Policy, 1999.

On the other hand, considering the inflow of foreign capital, technological innovation and R&D spending in the host countries by foreign companies and their branches represent a comparatively large proportion of the total industrial R&D investment in the host countries and the proportion has been increasing. However, the amounts in R&D spending in each country by the branches of foreign companies also greatly differ. Although the data are incomplete and those available may come from different times, we can still see the imbalance among different countries in Figure 5-4. For the manufacturing sector in countries such as Hungary and Ireland, the proportion of foreign investment in R&D is very high, but the proportion is very low in Japan, i.e., less than 1% (see Figure 5-4).

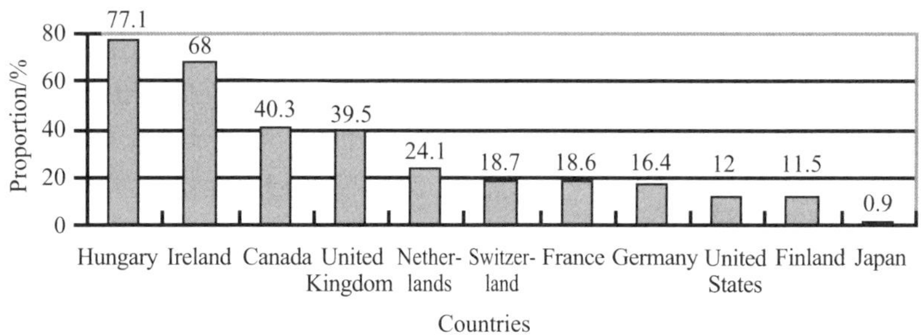

Figure 5-4 Proportion of foreign investment in different countries for R&D in the manufacturing sector

Note: Data for Ireland and Germany are for 1995; data for Hungary and Finland are for 1997; the other data are for 1996.

Source: OECD. Measuring globalisation: the role of multinationals in OECD economies. Paris: OECD, 1999.

Two-way investments among the big-triangle countries for internationalization of technological innovation are very distinct. However, since the 1990s, the geographic distribution of investment in technological innovation has gradually shown a trend of dispersion. On the one hand, developing countries, especially emerging industrialized countries, have become the emerging markets for investment in technological innovation and R&D activities by developed countries. For example, in the United States—the country with the largest overseas R&D investment—excluding its investment in Brazil which is the developing country that receives the largest amount of overseas R&D investment from it, in 1997, the total R&D spending in the 11 emerging markets, including Israel, Mexico, Taiwan of China, Hong Kong of China, Singapore, Argentina, R.O. Korea, Chinese mainland, Malaysia, South Africa and India, by overseas branches with a majority of its stock held by U.S. companies, exceeded $1.2 billion. Although the total investment in these emerging markets is very small, the momentum of development is very strong (see Table 5-1). At the same time, R&D

investment in the United States by the emerging markets has also been growing rapidly. R.O. Korea, Singapore, Israel, and other countries or regions have established—by holding stock—R&D branches in the United States (Dalton & Serapio, 1999). And, China has become the country with the most rapidly growing investment in R&D activities in the United States (Tang & Zhang, 2001).

Table 5-1 R&D spending by the United States in the emerging markets (100 million USD)

Market	1992	1997
Israel	0.24	2.09
Mexico	0.76	1.32
Taiwan of China	0.54	0.87
Hong Kong of China	0.13	0.84
Singapore	1.12	0.73
Argentina	0.15	0.43
R.O. Korea	0.14	0.42
Chinese mainland	0.03	0.35
Malaysia	0.12	0.32
South Africa	0.16	0.22
India	0.03	0.22

In addition to the imbalance among countries in inflow/outflow of capital investment for internationalization of technological innovation and R&D activities, there are differences among industrial sectors. Due to the differences in economic structure between the home country and the host country, there are differences among countries in the industry structure of overseas R&D spending. However, specific to a certain country, there are degrees of concentration on some industries for overseas R&D investment by multinationals. In 1997, the manufacturing industries with the highest degree of globalization for the United States were pharmaceuticals, automobile and auto parts, and then food and beverage; the R&D spending in the United States by the branch units of foreign companies was mainly concentrated in three sectors, namely pharmaceuticals and biotechnology, electronics, and industrial chemistry (Du, 2001).

(3) Intensity of R&D activities and internationalization

A study by Patel and Pavitt (1998) shows that apart from the pharmaceuticals industry, the proportion of innovative activities carried out domestically by an enterprise increases with the intensity of the technologies involved in its industry. The reasons are probably that products in the conventional industries and with a low level of technology content need adaptations to better meet the local tastes or that technological activities need to be done near the origin of raw materials. Thus, although the technological intensity in the industrial sectors is not high, the degree of internationalization in technological innovation and R&D activities is high, but high-tech products are universal worldwide. As a result, there is no need to set up overseas R&D units to carry out product improvements in order to meet the local needs; therefore, the degree of internationalization is low.

For multinationals, the intensity levels of their R&D activities (the proportion of R&D spending in overall sales) and the internationalization of their R&D activities (the proportion of overseas R&D spending in their total R&D spending) vary. By the intensities of R&D activities and R&D internationalization, multinationals can be divided into four categories and with economic development they will move in two directions (Figure 5-5). Considering the scale of economy in innovative activities, proximity to the company's headquarters, and strategic reasons including protecting high-level technological secrets in the company's headquarters and controlling key assets, some companies—although their overseas technological innovation and R&D activities are increasing—still carry out a majority of their innovative activities domestically (Fujitsu of Japan is an example for this type). Some other multinationals carry out their R&D activities globally, and with the cooperation in their global R&D activities and the integration of their R&D networks, their respective overseas R&D spending accounts for 90% of their respective overall R&D spending (ABB is an example for this type).

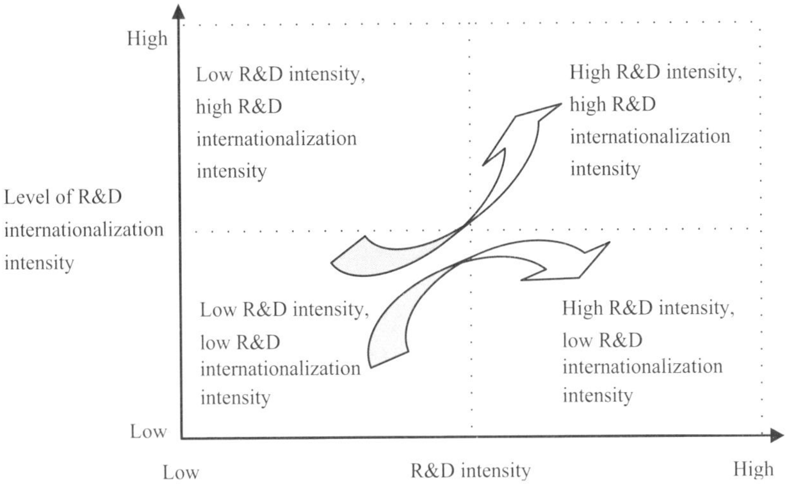

Figure 5-5 R&D intensities and R&D internationalization intensities in multinationals

(4) Models to direct the investment flow for international technological innovation

Du (2001) defined three models to direct the investment flow for overseas innovation and R&D activities—"upstream flow," "downstream flow," and "parallel flow" (Figure 5-6).

Internationalization of technological innovation and R&D featuring investment "upstream flow" tend to have investment flow from low-technology-level areas or countries to technologically advanced areas or countries. The objective of such international innovation activities is to track and acquire advanced technology in the technologically advanced areas or countries. Multinationals seek to establish innovative and R&D activities in places near technology centers (Davis, 2000). This model for international investment in technological innovation can represent not only investment in innovative activities in developed countries by developing ones but also investment between developed countries, since there are disparities among different advanced economies themselves in the technology level of any particular industry. Both Europe and the

United States invest in technological innovation in Japan's electronics industry only because the technology level of Japan's electronics industry is the highest in the world.

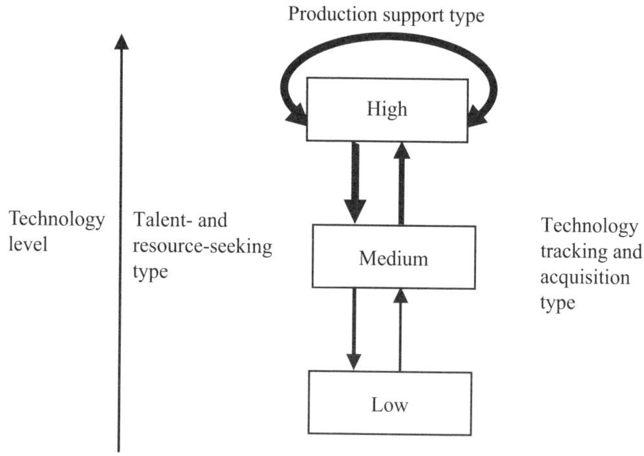

Figure 5-6 Investment flow for internationalization of technological innovation

Internationalization of technological innovation and R&D featuring "downstream flow" investment tend to have investment flow from technologically advanced areas to low-technology-level areas or countries. The objective of such innovation internationalization is to seek more talent, markets, and other resources. To most multinationals, the main goal for setting up R&D units overseas is to make adjustments to their products so as to make those products more suitable for the local market rather than to acquire foreign technologies. In fact, investment in international R&D more often flows from high-technology-level and high-R&D-intensity countries to low-technology-level and low-R&D-intensity countries and not the other way round (Guellec & Van Pottelsberghe de la Potterie, 2001). A majority of technological innovation and R&D activities carried out by developed countries in the developing ones are of this type.

Investment in internationalization of technological innovation and R&D of the production support type tends to feature "parallel flow," which, at the level of overall pattern, mainly takes the form of flow among the developed countries presently. However, in the case of one company, the direction of capital flow can go from developed countries to developing ones or from developing countries to developed ones, showing the nature of a "mixed flow."

5.1.2 Status quo of capital investment and allocation of China's internationalization of technological innovation

(1) Scale and source of capital investment in China's internationalization of technological innovation

In 2001, among the surveyed enterprises, of those with independent R&D branches overseas, 33% had annual spending on overseas R&D and technological innovation below 5 million yuan, 44% between 5 and 20 million yuan, 22% between 20 and 50 million yuan annually, and none spent more than 50 million yuan on overseas innovation. Compared to the spending on the overseas innovation and R&D activities by multinationals of other countries, such spending by Chinese enterprises is still very low. However, most enterprises indicated that they would increase the investment in overseas innovation activities. According to the survey data, by 2005, 66% of the enterprises estimated that their overseas investment in innovative activities would exceed 20 million yuan and half among the 66% would invest more than 50 million yuan in overseas innovative activities (Figure 5-7).

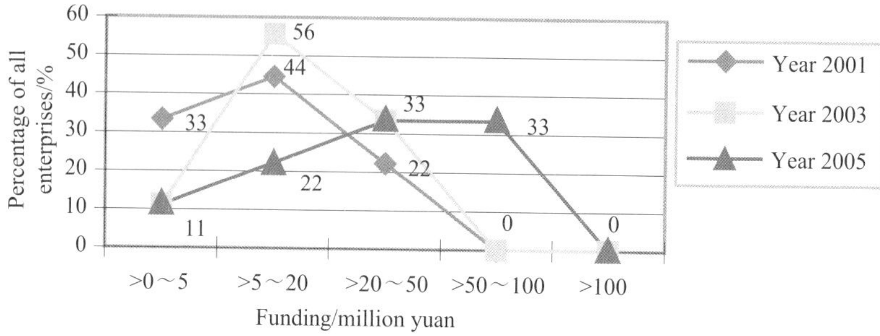

Figure 5-7 Size of funding for overseas R&D

Considering the funding sources for overseas branches, we included seven choices in our investigation: parent company, overseas subsidiary, the Chinese government, foreign partner of cooperative company (joint venture), domestic NGO, foreign NGO, and other sources. The result of the survey shows that those with 100% investment provided by parent company (including those with no independent overseas subsidiaries) account for 20.8% of the total number of the surveyed enterprises; those with 80% or more investment in overseas R&D branches provided by the parent company account for 75%; and all the overseas R&D branches have at least half of the investment provided by the parent company in them. Obviously, the main funding source for the overseas R&D branches of Chinese enterprises is the parent company. It is estimated that within the next five years, there will be significant changes to the sources of investment for overseas technological innovation and funding from sources other than parent companies will increase (Figure 5-8).

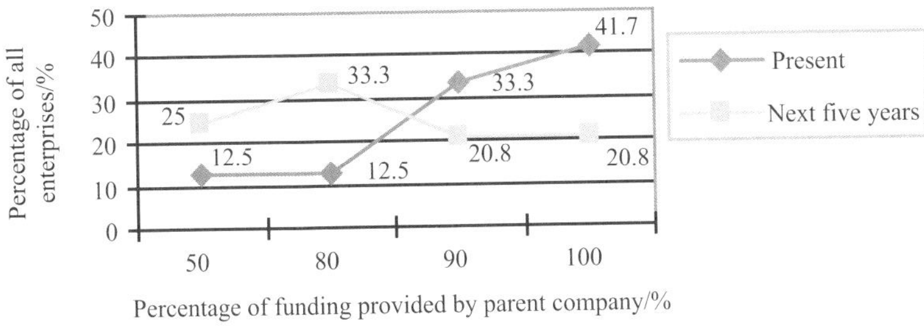

Figure 5-8 Sources of funding for overseas R&D units

(2) Models to direct investment flow for internationalization of technological innovation of China

Analysis of the surveyed enterprises shows that at present 40% of the overseas R&D branches established by Chinese enterprises are concentrated in the United States, 13% in Europe, and 7% in Japan. Another 13% are concentrated in Southeast Asia and 27% are distributed in other parts of the world—here, other parts of the world refers to some third world countries other than those in Southeast Asia. Within the next five years, the basic distribution structure of the overseas R&D branches will not change (Figure 5-9). This result shows that the investment flow pattern of technological innovation of China is mainly reflected in the "upstream flow" model, which includes setting up information window, tracking, monitoring, and learning advanced technologies from Europe, the United States, Japan, other developed countries and from other regions of the world. Gaining proximity to knowledge centers and establishing supplementary sources of expertise are the main strategic goals of internationalization of technological innovation by Chinese enterprises. Meanwhile, some Chinese enterprises have established R&D units in third world countries; the main purpose is to meet the needs of production support—they need to provide technical services and technical support for the production and sales in these regions and to adapt their products to the local markets. Such innovation internationalization based on production and market support can also flow to developed countries and regions, thus displaying the feature of "mixed flow."

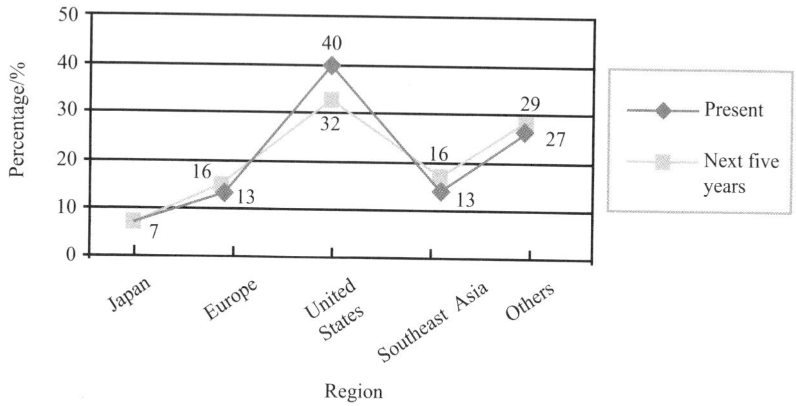

Figure 5-9 Distribution of overseas R&D branches by region

5.2 Allocation of human resources for internationalization of technological innovation

5.2.1 Selection and acquisition of human resources for internationalization of technological innovation

Information technology, especially the Internet, allows companies to carry out technological innovation and R&D activities in a wider area. The new reality that came about is that it takes people of different expertise to make a joint effort in order to creatively solve problems and invent new products. If the "human resources" element is not the only important factor in this era of knowledge economy, it is at least one of the important factors. People with a global mindset and a global behavioral pattern are the majority in globalization. For many companies, developing suitable personnel shows the true meaning of "global thinking."

(1) Analysis of motivation for recruiting overseas personnel

As the proportion of employees working in the overseas R&D subsidiaries and branches of multinationals sharply increases, the proportion of local R&D and management personnel from host countries working in internationalization of technological innovation for multinationals also increases sharply. Multinationals tend to recruit—in the host countries—a large number of the existing high-quality technical personnel, engineers, scientists, and experienced management personnel to engage in development in the high-tech industry and to work in enterprise decision-making; as a result, there is a great increase in the number of local scientific researchers and management personnel. The quantity and quality of human resources in the host countries, level of scientific research, and relevant scientific research infrastructure have become the key factors in attracting direct investment from the multinationals for R&D activities. The main reasons multinationals recruit local R&D and management personnel in the host countries for internationalization of technological innovation are presented as follows.

1) Taking possession of and getting close to the outstanding talent of the host country

At the core of every innovation, it is human resources that generate creative ideas (Von Zedtwitz, 2002). On the one hand, in today's environment where competition among enterprises is getting ever more intense, demand for high-quality innovative talent is getting increasingly strong and possession of outstanding innovative talent is crucial for multinationals in gaining advantages. On the other hand, the nurturing of talent takes a long cycle and the quantity and quality of supply are restricted by various factors; thus, shortage of R&D talent is a common challenge faced by many countries. Therefore, there is a need to recruit talent internationally in innovation internationalization to make up for the shortage of domestic talent.

Considering a company's characteristics, for technology-oriented type of growth companies and research institutes, it is especially important for them to be in possession of or to be close to outstanding human resources. In order to acquire new knowledge resources, such companies and research institutes must be located in places with a high density of research talent. For example, there are many informal gatherings of experts among the intellectual groups in Silicon Valley. Multinationals from Japan and Europe as well as from some other developing countries have rushed to the place one after another, and many of them have set up their R&D branches in the Silicon Valley area, making full use of its S&T human resources to advance their own innovative and R&D activities. Considering industry characteristics, pharmaceutical and biotechnology researches are more fundamental with new product development in these areas more closely linked to basic research, and thus, overseas R&D units usually need to be close to a scientific research base. The electronics industry is an intellectually intensive industry, and thus, overseas R&D institutions tend to be close to prestigious universities and research institutes so as to take full advantage of the research bases of the host countries, seek as well as expand on opportunities to cooperate with the local universities and research centers, and recruit suitable scientific research personnel. Considering national characteristics, acquiring high-quality talent is likely to be especially important for the headquarters of multinational companies in small countries. To some extent, they lack the technological talent needed even more for the domestic markets (Hakanson & Zander, 1988). For small countries like the Netherlands, Switzerland, Norway, and so on, they have a number of large companies, but these countries are small, with limited human resources, and thus, it is very difficult to domestically find various talent required for technological innovation. Their needs for global human resources are even more obvious.

In addition to directly recruiting innovative talent abroad, companies can also make use of local human resources overseas through various cooperation methods or other indirect means, such as approaching researchers at universities as well as other institutions. Collaboration with and closeness to the local scientists and engineers can significantly enhance the competitiveness of the company and create new and better products.

2) Obtaining human resources with lower cost than that in the home country

One of the benefits of internationalization of technological innovation is the ability to make use of the local technology bases. The branches of a company can acquire high-quality local experts who have already mastered the necessary skills and can be recruited and immediately start to work. Thus, the company can save the expenses of costly internal training. The abundant availability of good and trained human resources for technological innovation is a major driving force in the decision-making process of multinationals to move a portion of their R&D activities to a certain region or country in order to reduce the costs of global innovation and R&D activities (Kumar, 2001). Since the 1990s, multinationals have rushed into R&D in China one after another, with a main motivation to make use of China's abundant and cheap S&T talent. China is widely acknowledged to have the greatest number of the world's highly intelligent people, and the technological personnel generally have solid scientific training, good competency, and dedication to research activities. However, China's existing S&T system does not provide enough opportunities for the personnel to enhance their knowledge and intelligence. Under such circumstances, multinationals only need to put in much less effort in China than elsewhere to mobilize them and make use of their talent (Du, 2001). Technologically advanced countries or regions invest in technological innovation and R&D activities in low-technology-level countries, with the principal motivation to obtain relatively cheap but outstanding local human and other resources in the host countries.

3) Promoting exchange and communication in international innovation activities

On the one hand, the creativity in international innovative activities and the geographic dispersion of the sites of such activities make it extremely important to engage in information exchange and communication during international innovation. On the other hand, in the process of international innovation activities by a multinational, innovative and R&D ideas move from one individual scenario to another and move from a geographic location to another; to a certain extent, the process will very likely warp or filter the information conveyed. An international innovation team—inside the multinational—comprising people from different companies, different countries, and different cultures will develop its own special "language," subculture, and methods for exploring and analyzing the environment. Various differences among various countries in language, ethnic culture, custom, environment, etc. cause information about international innovation activities to get distorted in the delivery process. In order to reduce the degree of information asymmetry in the innovation activities conducted by multinationals, the innovative activities need to be localized to satisfy the local market and multinationals should also recruit local people in an effort to reduce the level of information asymmetry (Zhao et al., 2001). To ensure effective communication and coordination within a geographically dispersed internationalized innovation network and to maintain its creativity and high efficiency, effective human resource management is one of the key factors. Most crucial among the employees are outstanding project managers/leaders; appointment of the right team leaders can reduce communication costs and increase the degree of information security, network transparency, and knowledge credibility (Moenaert et al., 2000).

(2) Selection of project manager/leader for internationalized technological innovation activities

The selection of team leaders is very important for the success of an innovation project; management personnel and team must first select and train a project manager/leader. Although more and more companies have started to carry out cross-border projects, a majority of such projects lack a suitable project manager. With the development in internationalization, project management becomes more challenging, more unpredictable, and more uncertain; in addition, there will be more changes and surprises. This requires a professional project manager with excellent skills. If an inexperienced project leader is chosen, he/she will oftentimes lead the project into difficulty from the very beginning. Because the criterion used by many companies in choosing project managers is "the most easily found, not the most suitable," the appointed project managers are often deficient in knowledge and experience and thus not able to lead the process, and not knowing who, within the communication network of the multinational, are the potential collaborators. Therefore, they would fail to consult them, which will lead to a chaotic atmosphere within the organization.

R. K. Moenaert, F. Caeldries, A. Lievens, and E. Wauters (2000) and Maximilian Von Zedtwitz and Oliver Gassmann (2002) believed that a good project manager/leader should be able to process cross-functional and cross-subject information; should be able to maintain the continuity of the project, select necessary capabilities, adapt to different cultures, bridge cultural differences, control information flows, and harmonize the needs of different individuals; and should be able to correctly make use of the knowledge within a multinational about how to operate the entire product development process and to solve the problem of low efficiency that exists internally within the multinational.

(3) Selection of R&D personnel for internationalization of technological innovation

Even in the developed countries, nowadays, the value of output created by traditional industrial sectors falls far behind that created by high-tech industrial sectors; the transformation into a national knowledge economy becomes increasingly obvious; and the development, training, and use of knowledge-type talent will become a common theme in every country of the world. However, the shortage of R&D personnel will also be a challenge to be faced by all countries, and therefore, gaining proximity to and making use of outstanding R&D personnel locally in the host countries form an important strategy in internationalization of technological innovation, especially when it comes to high-tech R&D talent. The increasing demand for high-tech talent has triggered a more and more intense competition for the talent between foreign multinationals and host country's companies.

In addition to possessing exceptionally professional and technological capabilities, in order to ensure the effectiveness of global communication and coordination methods and to maintain the creativity and efficiency of the organization, technological and R&D personnel must be able to use modern telecommunications technologies; must have communication skills, strong adaptability, and willingness to learn; and must be able to work well in a team and work creatively.

5.2.2 Human resources for China's internationalization of technological innovation

(1) Scale of China's internationalization of technological innovation workforce

According to our survey, there are few researchers in the overseas R&D branches set up by Chinese enterprises: enterprises with more than 100 R&D personnel in their overseas R&D branches account for 21% of all the enterprises with overseas R&D branches and only 7.46% of all the surveyed enterprises. Among those with overseas R&D branches, more than half of them have no more than 20 R&D staff in those branches (Figure 5-10). This is mainly because the motivation of Chinese enterprises to set up overseas R&D branches is mostly technology monitoring and technology learning and is rarely technology transfer and technology development.

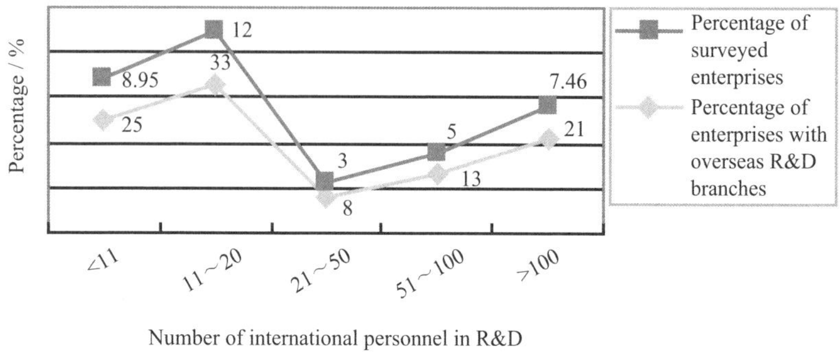

Figure 5-10 Scale of international personnel in R&D

(2) Sources of China's internationalized technological innovation workforce

When considering the employee structure of the overseas R&D branches, we noted that the proportion of foreigners in most of the overseas R&D branches controlled by Chinese enterprises is very low: 67% of the surveyed enterprises have less than 10% of foreigners among the researchers in their overseas R&D branches and in 55% of the surveyed enterprises there are basically only Chinese managers in their overseas R&D branches (Figure 5-11). A main reason is that many Chinese enterprises are not well known overseas, and thus, it is difficult to recruit suitable foreign researchers. The cost involved in using foreign researchers is also an important factor in limiting the use of foreign researchers by the surveyed enterprises. In this regard, recruited foreigners are mostly foreign Chinese; in addition to cost, culture and communication also influence the recruitment process. However, we have also noticed that there are individual cases among the surveyed enterprises that have achieved localization fairly well in their research and have fully tapped into the technological advantages of local researchers and the advantages of their familiarity with the local environment: Wanxiang Group is one such example. The proportion of local researchers and managers in the overseas branches of Wanxiang Group is very high (between 70% and 90%), and so far Wanxiang Group's localization strategy has been implemented successfully.

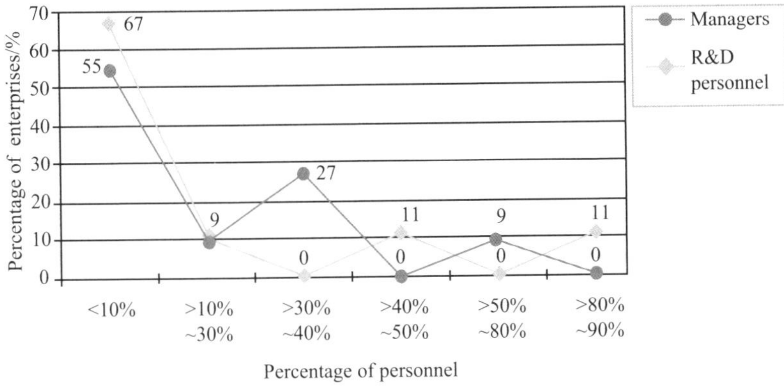

Figure 5-11 Proportional distribution of foreign R&D personnel and managers in overseas R&D branches

(3) Factors influencing personnel selection for China's internationalization of technological innovation

When selecting managers for overseas innovation projects, the surveyed enterprises generally believe that, for researchers of different countries, very high communication and coordination capabilities, good skills in exchange and communication with external people and R&D units, and access to a wide range of information are the most important factors to consider. Also, managers must be good at stimulating the enthusiasm and creativity in employees; fully understand the parent company's strategy and even find opportunities to participate in the parent company's technological strategy; be familiar with local social as well as cultural background; be good at exchange and communication with the R&D center of the parent company, convey the status of the local branch to the R&D center in a timely manner, and pass on the messages from the R&D center of the parent company to the local branch; and have a high level of technical competency and a wide range of knowledge. Companies generally believe that the cost involved in searching for managers is not a major factor in consideration and that the past experience in managing technological innovation is not very important either (Figure 5-12), which is basically in line with the abovementioned conclusions drawn by Moenaert et al. and Von Zedtwitz and Gassmann.

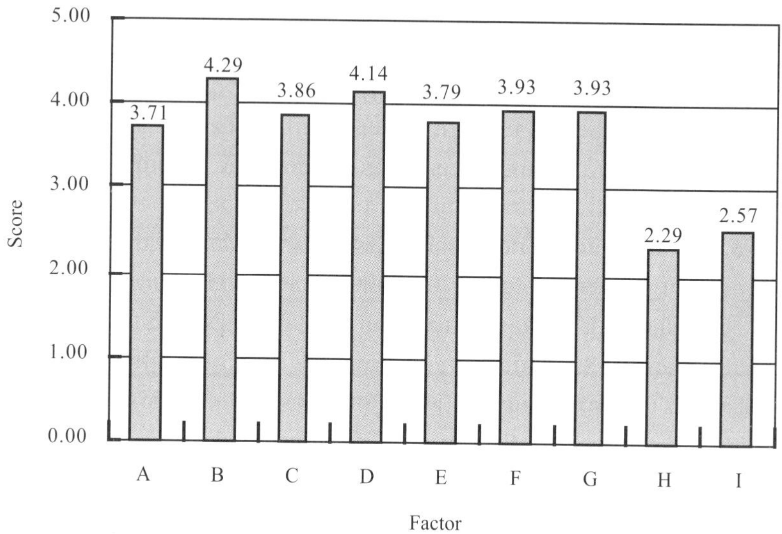

A: A high level of technological competency and a wide range of knowledge;

B: High communication and coordination capabilities of researchers from different nationalities;

C: Familiarity with the local social and cultural background;

D: Skills in exchanging and communicating with external people and R&D units, with an access to a wide range of information;

E: Skills in communicating with the R&D center of the parent company, conveying the status of local branch to the R&D center of the parent company in a timely manner, and passing various messages from the R&D center of the parent company to the local R&D branch;

F: Skills in stimulating enthusiasm and creativity in employees;

G: A full understanding of the parent company's technological strategy by even participating in the development and implementation of the parent company's technological strategy;

H: Cost incurred by the company in finding competent management personnel;

I: Candidate's past experience in managing R&D projects.

Figure 5-12 Factors for selecting overseas R&D project managers

When selecting researchers for overseas R&D branches, the surveyed enterprises attach special significance to qualities of the candidates, including dedication, level of technological expertise, abilities in international exchange and cooperation, and communication skills. In addition, they also give more consideration to other qualities of the candidates, such as ability to gather information, ability to learn, creativity, insight, and—for Chinese researchers—the ability to understand and know the social and cultural background of the host country (Figure 5-13). Obviously, in addition to the general requirements for scientific researchers in technological innovation, such as technological expertise, learning ability, creativity, insight, and dedication, there are special requirements in international exchange and communications for R&D personnel working for overseas R&D institutions.

The rapid increase in the amount of technological information and the ever-growing trend of globalization require us to give more attention to managing human resources. High-quality talent is at the core of each activity of any innovation, but this is also the weakest area of Chinese enterprises. The acute talent shortage will directly impact the development of Chinese enterprises and internationalization of their technological innovation. In addition, human resources for technological innovation in Chinese enterprises have not yet achieved a true level of internationalization and localization; most overseas innovative researchers are Chinese nationals or overseas Chinese, and foreigners with other-than-Chinese cultural background account for a very small proportion. The ability to strengthen the cultivation of international researchers and management personnel, optimally use the human resources worldwide, and thereby improve their own technological capabilities will directly impact the process of technological innovation and R&D internationalization by Chinese enterprises.

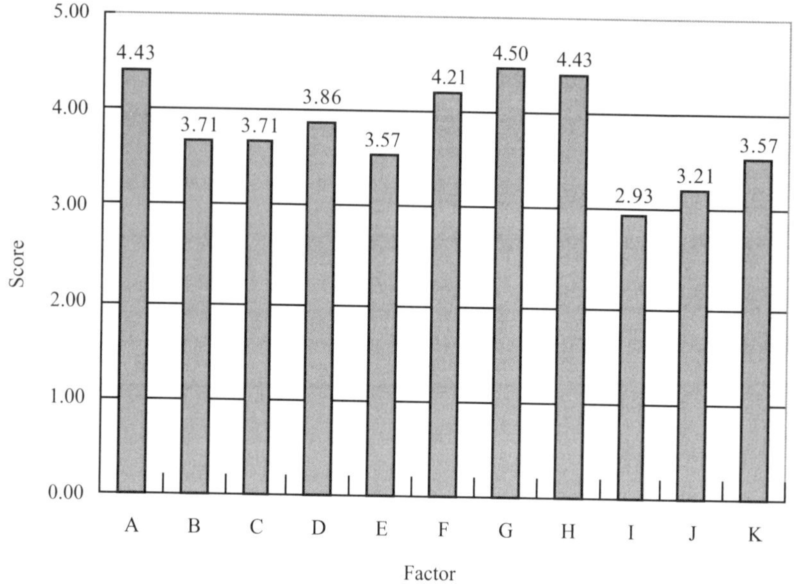

A: Technological competency of researchers (including Chinese and foreign researchers, the same as for all the factors given below);

B: Learning ability;

C: Creativity;

D: Ability to collect information;

E: Insight;

F: Communication skills, especially skills in communicating with researchers of different nationalities;

G: Dedication;

H: Skills in international cooperation and exchange;

I: Language skills;

J: Knowledge of foreign researchers about Chinese cultural and social customs;

K: Knowledge of and familiarity with the host country's social and cultural background in case of the Chinese researchers.

Figure 5-13 Factors for selecting R&D personnel in overseas branches

5.3 Acquisition, development, and transfer of technological information sources for internationalization of technological innovation

5.3.1 Acquisition, development, and transfer of technological information sources

The ability to acquire innovative ideas and to quickly and economically identify innovation opportunities is the prerequisite for the success of an enterprise's technological innovation. R&D and technological innovation require that information be communicated, transmitted, and exchanged in all directions inside the enterprise and to the external parties so as to nurture innovative ideas. A research has shown that of 100 new product ideas, only 39 can enter product development process, 17 can come out of product development process, 8 can truly enter the market, and only 1 can ultimately achieve business goals (Chen, 2001). Effective R&D and technological innovation require systematic search for new ideas and new sources of information.

(1) Acquisition and development of information sources for internationalization of technological innovation

Globalization brings new knowledge, and multinationals have more channels and opportunities to acquire information and knowledge sources; thus, understanding the trends of overseas technological developments and enhancing the input of new ideas have become the starting point of carrying out overseas R&D and technological innovation for multinationals. Many multinationals continually set up knowledge absorption centers in different regions of the globe, so as to intensify its global search in relevant fields of research and technology. Specifically, international technological information is sourced from internal sections and external customers, competitors and suppliers.

Many innovative ideas come from within the enterprise. The internal sources include researchers in overseas R&D branches and in the R&D center of the parent company; enterprises can also find new ideas through formal investigative activities. Sales staff in an enterprise, especially those of the overseas marketing department—because they frequently contact the users of their products—can come up with some new product ideas. Other departments of an enterprise can also contribute to the generation of new ideas.

For a market-oriented enterprise, information from the market is more important than any other information because market information reflects people's understanding of goods and services (Leonard-Barton, 2000). On the one hand, internationalization of technology always goes after the internationalization of production, research done abroad is often not in the intensive industrial sectors, and the emphasis of such research is to adapt products to local markets (Cantwell & Kosmopoulou, 2001). On the other hand, many multinationals first consider carrying out innovative activities in a dynamic and forward-looking market and their aim is, through setting up local R&D, to learn from the leading markets and satisfy the needs of upscale customers, so as to improve their own innovative processes (Gerybadze & Reger,

1999). Market and customer information is an important source of the creativity and is necessary for technological innovation; overseas branches of multinationals are often intimately familiar with local as well as global customers and market conditions, and thus can gather valuable ideas for research.

Competitors are another good source of creativity for technological innovation. Enterprises should launch a similar product to the market ahead of competitors and position its own new product development after analyzing the competitors' products. Information related to competition is crucial for successful R&D because it can trigger innovative ideas. For example, Ford's Jaguar uses Audi's accelerator pedal "tentacles," Toyota Supra's fuel gauge, BMW 528e's tires and a jack storage system, and 400 other similar features.

In addition, local cooperative enterprises, suppliers, universities, or laboratories in the host country can be the sources of technological knowledge and can produce creative ideas for technological innovation. New product development ideas from employees, suppliers, and competitors are often easy to implement; although different enterprises differ as to how to implement such ideas, they all know the importance of them. However, many enterprises do not limit the sources of creative ideas and gain new product ideas from all business partners. Leading enterprises integrate creative ideas from all sides: customers, employees, competitors, resellers, general distributors, suppliers, and other related parties (Golder, 2000).

(2) Sources of technological information for international innovation and forms of international R&D organizations

In internationalization of technological innovation, there are important relationships between the creativity source of overseas R&D and the organizational form of overseas R&D branches. Jorge Niosi and Benoît Godin (1999) conducted a detailed investigation on 18 Canadian multinationals and found that for overseas R&D branches with different organizational forms, the purposes and tasks are different and, thus, the creativity sources for international R&D are very different.

Jorge Niosi and Benoît Godin divided overseas R&D institutions into three categories. The first category corresponds to the related diversification strategy. The companies of this category are mainly in industrial sectors, such as machinery, transportation equipment, and metal products, and their overseas R&D units develop different types of related products. Each overseas R&D institution operates for the local market as well as for the global market and simultaneously demonstrates the characteristics of a domestic technological unit and a global technological unit. Technically, branch units enjoy a high level of autonomy in human resource management and project management, and the parent company hardly participates in project-related decision-making. The creative ideas for the R&D and technological innovation in such overseas R&D branches mainly come from local universities or public laboratories and sometimes from foreign sales departments, but rarely from the parent company.

The second category includes overseas R&D branch units of vertically integrated international companies. The companies are mainly in metal products, forest products, and chemical industry. They carry out the preliminary or basic research in the home country and conduct a more market-oriented overseas R&D research close to the market. Such overseas R&D units also have a high degree of autonomy, although not as high as that for branch units in the first category; their creative ideas for R&D and technological innovation mainly come from the overseas sales department and sometimes from local universities and public laboratories or the parent company.

The third category includes the overseas R&D branch units of the multinationals that have truly implemented a globalization strategy; such multinationals carry out R&D activities in different countries, developing a variety of related products. The main task of overseas branch units includes outward technology transfer for supporting the parent company's marketing activities. The creative ideas for international R&D and innovative activities of such companies are usually from the parent company and oftentimes from foreign marketing departments, but rarely from local universities or public laboratories in the host countries.

(3) Effective knowledge absorption capability

Existing knowledge and information are not sufficient to understand, absorb and transfer new knowledge and thus are not capable of generating new concepts or new ideas. The sources of technological innovation must be acquired and developed, but attention must also be paid to the flow of knowledge and information inside the company. The ability to transfer knowledge and information more effectively than that of the market is the reason for the existence of multinationals (Kogut & Zander, 1993), and the most important factor affecting the internal knowledge transfer is the ability to absorb knowledge (Lane & Lubatkin, 1998; Gupta & Govindarajan, 2000). Absorption capability is referred to as "recognizing the value of external information, absorbing it, and ultimately putting it to commercial use" (Cohen & Levinthal, 1990). The absorption capabilities of overseas branches differ, which will affect the effectiveness of knowledge transfer.

Minvaeva et al. (2000) believed that there are two dimensions to the absorption capability: on the one hand, employees have the ability to recognize the value of new information, to absorb it, and to commercialize it; on the other hand, employees have the motivation to acquire knowledge. Both ability without motivation and motivation without ability will lead to low performance.

5.3.2 Analyzing the present state of technological information sources of China for internationalization of technological innovation

The rapid development of information technology is a popular innovation, especially for developing countries that had little access to information resources in the past. In the process of the internationalization of Chinese technological innovation, the first problem to be solved is how to acquire innovative ideas, innovative information, and technology. The need to establish a set of methods to acquire and absorb foreign technology is urgent. Thus, for the vast majority of Chinese enterprises, technical search, technology monitoring, and project-based technology learning are the main strategic reasons that drive them for internationalization. We can have a look at Lenovo and Haier. There is no doubt that the fundamental purpose of Lenovo to establish the R&D center in Silicon Valley is to obtain the latest technology and get feedback about the business trends. However, for Haier, one of the leading enterprises in China with innovation internationalization, which has already been conducting R&D activities for technology transfer and technology support aimed at overseas markets (mainly the markets of less developed countries), the establishment of technological monitoring network is still in the preliminary stage for internationalization of innovation. Although Haier Group's international innovation has already gone through phases including joint venture, introducing foreign technologies with joint capital—establishment of its own technology research center and development system—technology export, and establishment of overseas R&D branches, it is still setting up technology monitoring points in multiple countries of the world in order to monitor the technological trends of competitors and the directions of technological development in its fields and to provide information for their localization effort.

The analyses of our questionnaires and interviews show that the main sources of ideas for overseas R&D and technological innovation of Chinese enterprises are overseas competitors, namely, local universities and research institutes in the host countries, R&D units in the parent companies, and local marketing department of the overseas units (Figure 5-14). These sources have been emphasized relatively more, and it is believed that their importance will stay at the same level or even go up in the future (Figure 5-15). However, overall, Chinese enterprises do not attach adequate importance to making use of a wider range of information sources and finding innovative opportunities more rapidly and more economically.

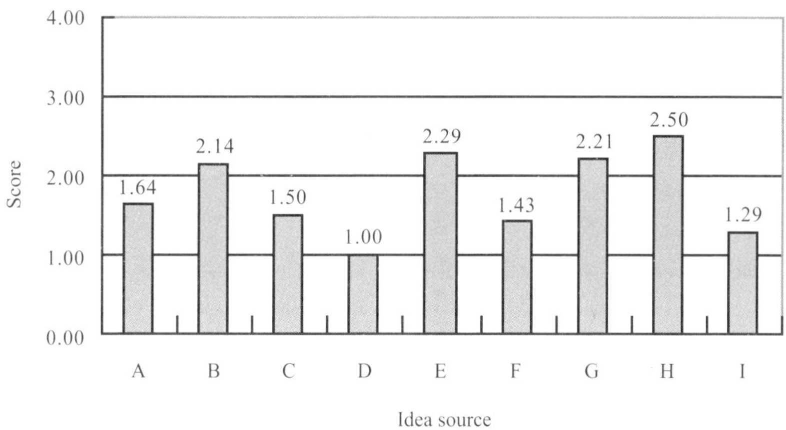

A: Local customers of overseas R&D unit in the same area;

B: Overseas marketing department;

C: Overseas production department;

D: Local companies controlled through M&As;

E: R&D department of parent company;

F: Other departments of parent company;

G: Local universities and research institutes overseas;

H: Overseas competitors;

I: Various periodicals, planned by the local government at overseas site.

Figure 5-14 Idea sources for overseas R&D

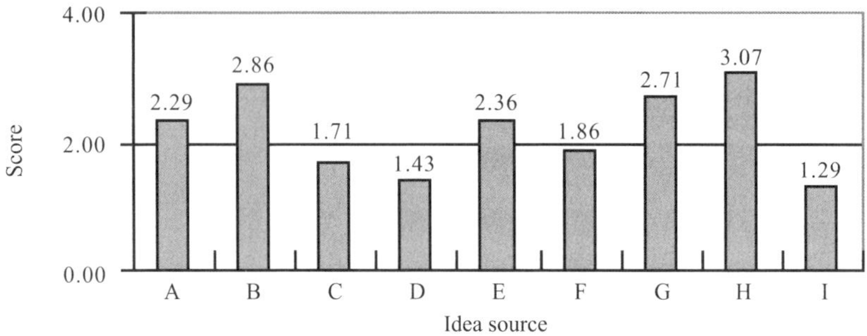

A: Local customers of overseas R&D unit in the same area;

B: Overseas marketing department;

C: Overseas production department;

D: Local companies controlled through M&As;

E: R&D department of parent company;

F: Other departments of parent company;

G: Local universities and research institutes overseas;

H: Overseas competitors;

I: Various periodicals, planned by the local government at overseas site.

Figure 5-15 Degree of importance of idea sources for overseas R&D

5.4 Effective resource allocation and integration in the process of internationalization of technological innovation for Chinese enterprises

5.4.1 Comprehensive analysis of models for directing innovation resource flow in the initial stage of internationalization

(1) Model for directing innovation resource flow in the initial stage of internationalization of technological innovation

Internationalization of Chinese technological innovation is a relatively recent development, the contact of most enterprises with the outside world is limited to jointly setting up R&D alliances with foreign enterprises, and very few enterprises have established overseas R&D branch units. Even for those R&D branches that had already been set up by Chinese enterprises, there are common problems such as limited investment in personnel as well as capital and inadequate R&D capabilities. China is still in the initial stage of internationalization of technological innovation.

In the initial stage of internationalization of technological innovation, Chinese enterprises mostly set up overseas R&D branches in the technologically advanced countries or regions, and the purposes of overseas R&D branches are mainly technology search as well as monitoring and technology learning that is project-based. Such enterprises are at a low degree of internationalization in production and marketing, and they have very limited number of products being sold in the overseas markets. In this mode, an enterprise's strategic orientation is very clear, that is, to carry out technological innovation activities for the domestic market and occupy the domestic market. China's domestic market is big enough for this mode, what's important for an enterprise is the domestic market, rather than the overseas markets. However, such enterprises have also realized that China's accession to the WTO has brought challenges and competition to them and that, by relying on the available technology and knowledge domestically, they cannot respond adequately to the competition from overseas manufacturers. Therefore, such enterprises need to establish overseas technology monitoring points or R&D branches, learn about foreign technology and market trends, relay technological information and other relevant information to domestic R&D centers, and plan as well as arrange R&D projects in domestic R&D centers. In this way, they can improve their potential technological innovation capabilities and enhance their ability to compete in the domestic market.

Considering the direction of resource flow, an enterprise in the initial stage of internationalization of technological innovation goes through the domestic headquarters to invest manpower and capital in the overseas R&D branches. The overseas R&D branches, via relevant activities, acquire information and knowledge about advanced technologies and relay useful information to the company's headquarters and related R&D center (see Figure 5-16) to provide new ideas for R&D and technological innovation, so that the company's products and technology can keep up with those of the developed countries. Thus, the enterprise will be able to occupy a favorable position in the domestic market.

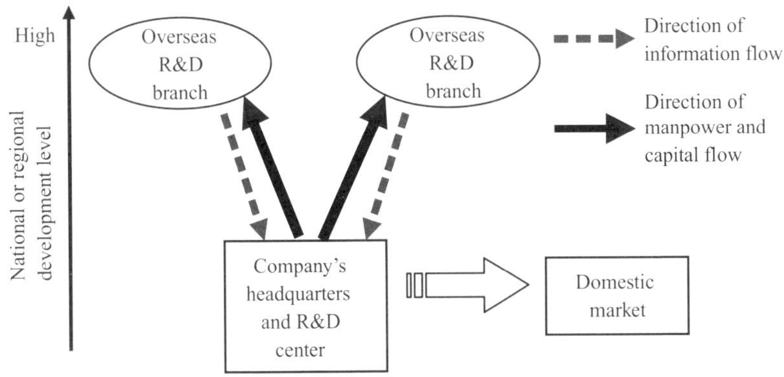

Figure 5-16 Model I for resource flow of the internationalization of technological innovation

---

Case study: Lenovo's internationalization of technological innovation

Compared to Lenovo's relatively complete international operations, its internationalization of technological innovation process started relatively late and the process has been in slow development. Broadly, the main driving factors of Lenovo's internationalization of technological innovation are as follows: (1) to bring the world's truly outstanding products to China and achieve localization of technology and technological processes, so as to carry out technology transfer, assimilate imported technologies, and become the market-leading brand in China's IT industry; (2) to keep up with the latest technological development in the industry, understand the needs of the international market, build up an international large-scale manufacturing industry, and provide production and design services for companies with international brand names.

With such strategic goals, Lenovo has set up R&D centers in Beijing, Hong Kong, and Shenzhen in China, and Silicon Valley in the United States. These centers have formed a system of specialization and cooperation: the center in Silicon Valley is an information station, constantly monitoring the latest technological and commercial trends in the world's largest computer market and providing timely feedback about important information; the center in Hong Kong carries out the development and pilot test of new products, such as motherboards; the center in Shenzhen directly supports the Hong Kong center—it is mainly established for saving personnel costs; innovative activities are mostly concentrated in China. The R&D center in Beijing is where experts from the Institute of Computing Technology of the Chinese Academy of Science—which represents the highest level of China's computer technology—are gathered. The research center in Beijing has multiple functions: the first one is to use its powerful quality evaluation expertise to carry out scientific tests and examination on new products developed in Hong Kong and then to provide test feedback to the Hong Kong center for product improvements, modification, and perfection; the second one is to develop various models of computers based on the motherboards developed in Hong Kong; and the third one is to develop software, such as the Lenovo Chinese character system, and to provide technology support for large-scale system integration projects (see Figure 5-17).

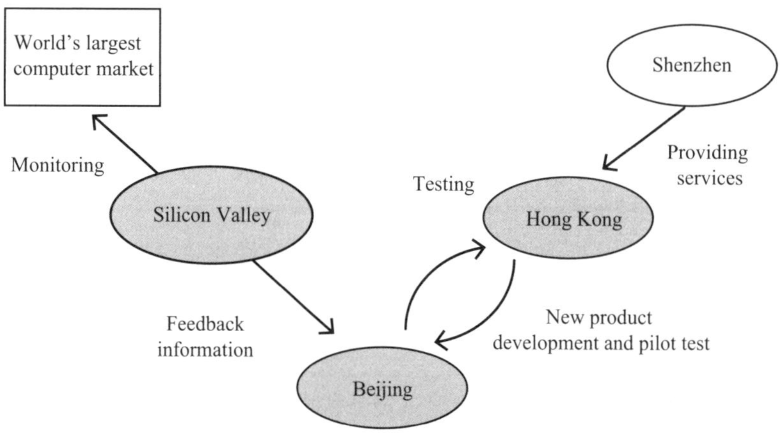

Figure 5-17 Lenovo's international R&D system

As shown in Figure 5-17, Lenovo is presently in the beginning stage of internationalization of technological innovation, its R&D activities are mainly carried out in the home country, its R&D unit in Silicon Valley is mainly for monitoring the world's advanced S&T, and thus it does not depend on overseas technology or market. Lenovo with the backing of the technological power of the Institute of Computing Technology of the Chinese Academy of Science—through the R&D systems in Beijing and Hong Kong in China and Silicon Valley in the United States—has established its product development capabilities that are adapted to changes in the international market. Meanwhile, by the overall allocation of a small workforce in Silicon Valley, a medium workforce in Hong Kong, and a large workforce in Beijing, Lenovo has achieved an R&D cost level that is significantly lower than that of computer companies in the advanced economies of Europe and the United States, and even lower than the cost level of computer companies in some other regions in Asia, and thereby has gained a relatively stable advantage in global competition.

The future Lenovo will be a "high-tech, service-oriented, and internationalized Lenovo." It will go from being the main brand in the IT industry in the Chinese market, through building large-scale international manufacturing capabilities, providing product manufacturing and design for companies with an international brand name, and thus accumulating strength, to eventually become an international brand. Lenovo is gradually moving towards greater internationalization, but currently it is still in the initial stage of internationalization of technological innovation. Becoming a truly internationalized company cannot be accomplished overnight, as it takes more than cosmetic surgeries, fundamental changes are necessary.

---

(2) Model for directing innovation resource flow in the development stage of internationalization of technological innovation

Huawei, ZTE, Potevio, Hisense, Haier, and several other leading companies have simultaneously entered the second stage of internationalization of technological innovation. They have not only set up overseas technology monitoring points but also more actively established R&D branches through direct investment or M&As. Their overseas R&D activities are not limited to technology monitoring and learning, as they also carry out technology transfer and technology support for locally based overseas production and marketing activities.

The manufacturing and marketing of the abovementioned enterprises have already achieved partial internationalization. The domestic market is their most important market, but because of the needs of medium- and long-term development, they attach great significance to exploring overseas markets since the overseas market will be the key factor for their further development. At present, their overseas markets are mainly in the less developed countries; some of these enterprises target some markets in the developed countries, exploiting China's low costs to gain advantage in those markets. The degree of innovation internationalization by them when compared to most other enterprises is much more advanced, the scale is larger, and the system is more robust. For resource acquisition and allocation, the means of resource acquisition for overseas R&D branches are no longer limited to the inputs mainly from the parent companies, the local recruitment and interunit personnel flow have greatly increased, and the sources of capital input have begun to diversify—the company's headquarters, overseas subsidiaries, venture capital, and various noncommercial organizations can become investing parties. Information exchange has changed the flow from a unidirectional one between the layers of organizational hierarchy to a bidirectional one, thus leading to the existence of information exchange among the branch units (see Figure 5-18).

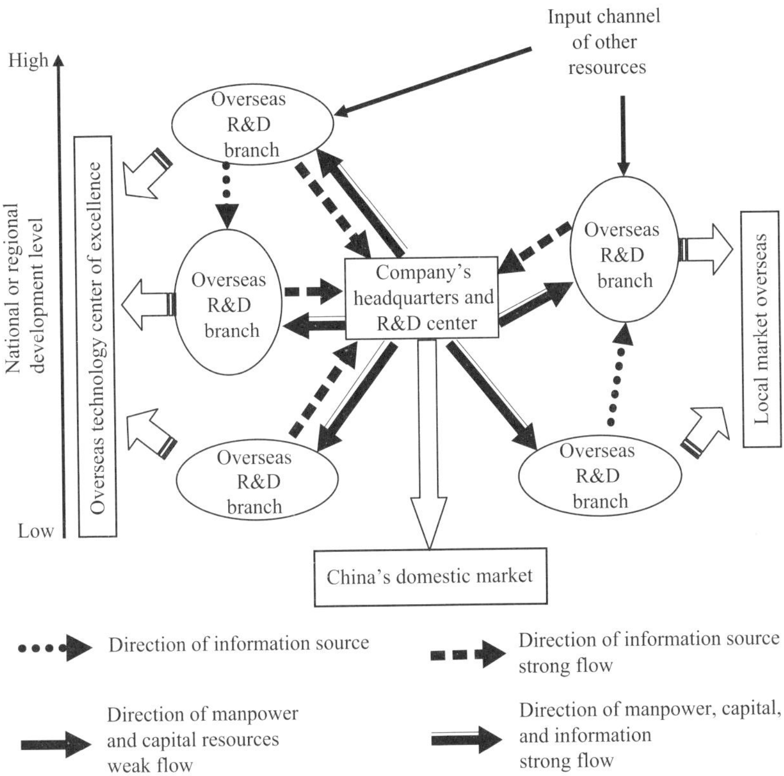

Figure 5-18 Model II for resource flow of internationalization of technological innovation

(3) Model for directing innovation resource flow in the advanced stage of internationalization of technological innovation

Enterprises that are in the advanced stage of internationalization of technological innovation truly carry out globalized technology innovation. These enterprises are usually large in size with tremendous strength, and their technological innovation is aimed at the international market and international technology. They invest directly in overseas R&D branch establishment, overseas local R&D personnel recruitment, and overseas R&D activities implementation. The status of the domestic R&D units is weakened, while the independence of the various R&D units is enhanced, and all the R&D units and the domestic R&D center form a relatively unified and coordinated network organization, in which knowledge and technology can be shared. Not only do some overseas R&D units transfer technological knowledge and information to the parent company, but also the parent company's headquarters and R&D center have strengthened their effort to transfer technological knowledge and information to the overseas R&D branches, and from there, the transfer of information and knowledge among overseas R&D branches has also been gradually strengthened. Considering the manpower and capital resources, their input in overseas units has increased. The increase in input not only includes resources from the company's headquarters but also those from a number of sources other than the company's headquarters (Figure 5-19).

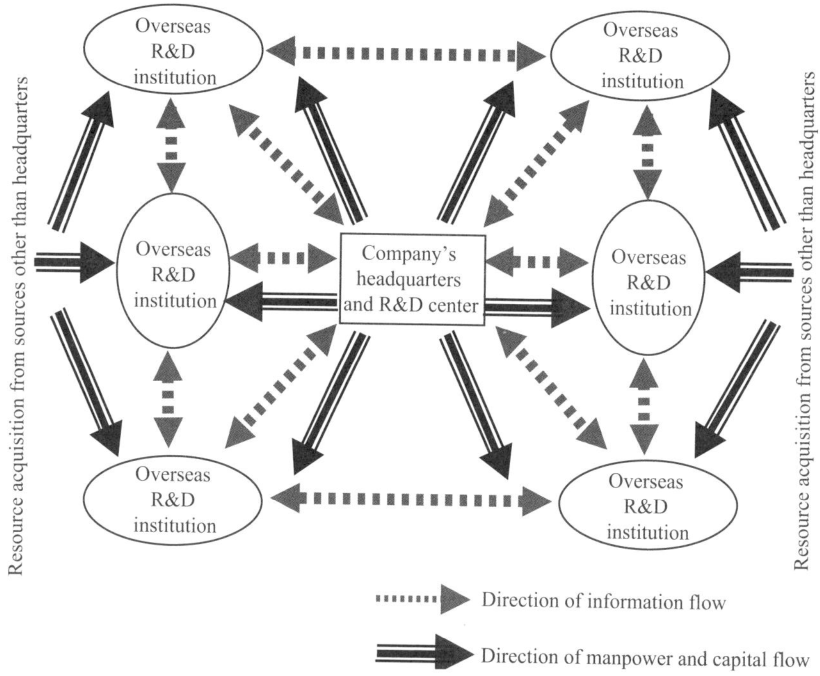

Figure 5-19 Model III for resource flow of internationalization of technological innovation

5.4.2 Effective allocation and integration of innovation resources in China's internationalization of technological innovation

Generally, China's overseas R&D and innovation personnel are mainly either overseas Chinese or those from the home country—foreigners from other cultural backgrounds account for a very small proportion. Capital investment is mainly from the parent company and that from overseas subsidiaries, the government, foreign cooperative joint ventures, various domestic NGOs, various foreign NGOs, and so on is very limited—zero for some enterprises. Information and knowledge flow is usually unidirectional, as it is the overseas branches that provide the technology and market information obtained overseas to the parent company. While China's international activities have been gradually shifting from technology monitoring and technology learning to technology transfer, with the improvement in each enterprise's capabilities in international operations and in R&D as well as innovation, enterprises will gradually become more focused on carrying out international technological innovation activities for the purpose of expanding and strengthening their potential technological capabilities. However, for a long period in the future, technology search and project-based technology learning will still be the main strategic motivation for innovation internationalization by Chinese enterprises, but R&D activities for technology transfer to overseas markets and for technology support in overseas markets will increase, which will be the mainstream of innovation internationalization by Chinese enterprises. In this process, the optimal allocation and integration of worldwide production factors will enable Chinese enterprises to make full use of the existing technological resources of various countries across the world and to gain the greatest benefit.

(1) Effective acquisition and allocation of funds for internationalization of technological innovation

Abundant capital, advanced technology and equipment, full-featured and complete research infrastructure, and so on form the foundation on which multinationals can carry out internationalization of technological innovation. However, small capital investment and singularity of investment source are presently important characteristics of resource allocation for innovation internationalization by Chinese enterprises; such relative lack of funds will directly affect the course of Chinese innovation internationalization.

1) Increasing the scale of investment for internationalization of technological innovation

Strengthening technological research by enterprises and constantly developing and providing innovative products with high technology and high value-added content have become the important means for monopolistic capital to gain high profits, seize the world market, and enhance competitiveness (Lin & Chai, 1998). In order to maintain and expand the technological advantages in their products, which are the most advanced worldwide in the fiercely competitive high-tech field, many large multinationals are willing to invest heavily in technological innovation research activities that are mainly for new products, new technological processes, and new materials. With a different scenario compared to the multinationals of developed countries, internationalization of technological innovation by Chinese enterprises is focused on technology monitoring, technology learning, and product development and improvements that are based on manufacturing and market support; thus, the demand for funds is relatively low. However, considering the current situation of investment in international innovation activities carried out by Chinese enterprises, the size of investment is still too small. The enterprises must increase investment in overseas R&D and innovative activities, especially enterprises that are already in the stage of making technology transfers to develop new markets. With an increase in the degree of innovation internationalization, capital input in overseas R&D and innovation will continue to grow.

2) Expanding channels of investment in internationalization of technological innovation

The singularity of investment channel for internationalization of technological innovation is another problem faced by Chinese enterprises in the internationalization process. This problem in turn leads to the relatively small scale of investment in international innovation activities. An important issue being faced by these enterprises in the innovation internationalization process is how to go beyond parent companies and expand the funding sources for overseas innovative activities. Most enterprises are aware of this point. According to our studies, in the next five years, the capital structure of overseas innovative activities will undergo big changes and investment from the parent companies will significantly decrease. However, it is expected that, on the whole, within next five years, funds from the parent companies will still account for a large portion and other channels, especially external funding sources of the enterprises, are rarely being considered by Chinese enterprises. As the enterprises increase the degree of internationalization of technological innovation, they should further expand the channels of investment.

Generally, the investment scale in China's internationalization of technological innovation is very small, and there is mostly only one investment channel. Aside from that of the parent company, there have been few investment channels of overseas subsidiaries, government, overseas cooperative (joint) companies, various domestic NGOs, and various overseas NGOs and the situation where investment is made truly by international and diversified channels has not happened. In the case of tight funding, how to raise R&D funds through multiple channels in the future, effectively support the development of innovation internationalization, and move toward internationalized and multichannel development is an important question to consider for

China's internationalization of technological innovation and is an important focal point of scholarly research.

(2) Effective allocation and integration of human resources in internationalization of technological innovation

Innovation talent will never be surplus. In the R&D activities and technological innovation of an enterprise, the importance of talent is becoming ever more obvious and the introduction, use, and motivation of talent are the most important parts of enterprise management in the future. The internationalization of technological innovation needs to attract and retain the best talent and to improve the ability, quality, team spirit, and innovative skills of employees. Talent acquisition strategy is the main feature of R&D and innovative activities carried out by multinationals. While enterprises invest large sums of money in overseas innovative activities, special attention should be given to the allocation and use of human resources.

However, at present, the number of Chinese innovative talent is small, and the level of their expertise is relatively low. The acute talent shortage will directly impact the development of Chinese enterprises and internationalization of their technological innovation. The following points are key to tackling this problem.

1) Full access to and possession of the host country's outstanding talent

Currently, innovative personnel in China's internationalization are mostly from the parent companies, and even employees of foreign nationalities are often overseas Chinese; one important reason for this is that the costs for acquiring outstanding human resources are too high for Chinese enterprises that also lack international brand recognition, and thus, it is hard for them to attract high-level talent. However, despite the presence of these difficulties, Chinese enterprises not only need to redouble their effort in fully developing and effectively allocating existing personnel but also must greatly increase the number of outstanding talent they recruit in the host countries.

For most Chinese enterprises, gaining access to foreign talent is perhaps a smart transitional approach. This approach includes cooperation with foreign universities and companies, participation in international forums, involvement in international research activities, and overseas technology training.

2) Emphasis on manager/leader selection for overseas R&D and innovative activities

Cross-border projects require highly skilled professional project managers. A good scientist is often not an excellent project manager; technological expertise alone is far from enough for cross-border projects. Many Chinese enterprises have not recognized the importance of project managers in internationalization of technological innovation.

A project manager is fully responsible for the outcome of the project and is also responsible for the progress in the technological aspect. With the ever-increasing environmental uncertainties, the need for scarce resources will continuously increase and companies are required to closely cooperate on global projects: this not only means that R&D activities must be carried out in sync but also requires that functional integration be achieved—R&D, innovation, and other functions need to be integrated into one process (Von Zedtwitz, 2002). Thus, the key role of project managers in internationalization of technological innovation is to facilitate coordination, communication, and integration among the many cross-functional, cross-disciplinary, and cross-border innovative activities inside a multinational. In these aspects, they play the role of bridges (Figure 5-20); in the early phase of a project, such communication and coordination methods are especially important.

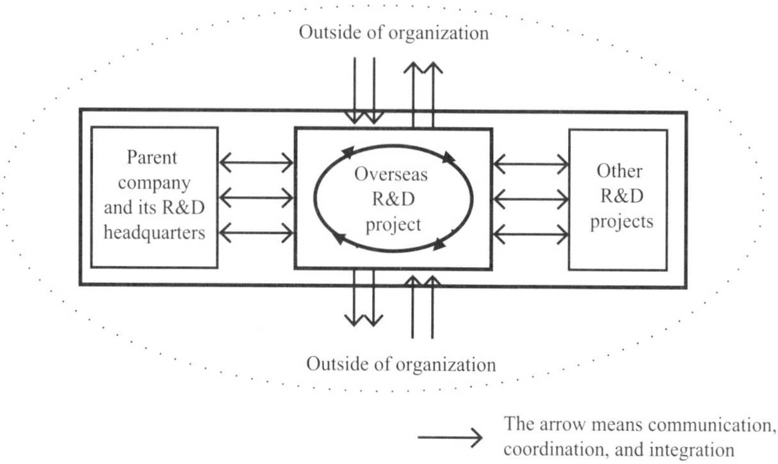

Figure 5-20 Communication and coordination by the project managers of overseas innovative activities

(3) Effective access to and transfer of technological information

To change the status quo of low-level technological capabilities and to strive to gain advanced technology and experience from abroad are the key steps for Chinese enterprises to enter the world stage after China's accession to the WTO. An open world has given China and other developing countries an excellent opportunity to obtain the necessary information and knowledge resources, which will be the idea source for tomorrow's technological innovation. However, presently, most enterprises have not been making good use of the extensive information resources overseas, and their exchange and communication have not been effective. Better use of more extensive information resources and faster and more economical discovery of innovative opportunities form the foundation for the success of internationalization of technological innovation by Chinese enterprises.

1) Expanding the channels of technological information

Technological information is the source of new ideas for R&D

and technological innovation. Advancement in information technology has brought opportunities for developing countries such as China to access technological information of the world's advanced technology. Chinese enterprises—in the process of innovation internationalization—have not yet learned how to effectively make use of the wide-ranging information sources. Aside from the conventional sources of innovative ideas, such as some sources inside the enterprise and some sources outside the enterprise, namely, the market, customers, suppliers, vendors, and competitors, Chinese enterprises should pay attention to some low-cost technological information sources that are better suitable for the current situation in China. First of all, the enterprises can access information through public information sources; this is the most convenient and inexpensive approach, though it has not been fully utilized (Wu, 2000). They should set up appropriate channels and assign specialized personnel for collecting, retrieving, and analyzing domestic and public information sources that can be exploited. Second, the enterprises should strengthen the cooperation between domestic and overseas universities and research institutes and actively participate in various academic exchanges and seminars, in order to access relatively advanced technological information. Third, Chinese enterprises can make use of reverse engineering technology to map and analyze the existing products and equipment so as to acquire technological information. However, further R&D on the product and equipment should be carried out, and infringement of intellectual property rights should be avoided.

2) Building up more channels for transmission and flow of information resources

After a multinational recognizes the importance of some internal or external information or knowledge and then absorbs it, the effective transmission and exchange of such information and knowledge resources become important. However, a multinational company is a "decentralized network," which may produce or obtain information and knowledge resources in one place and use such resources in another place. Information and knowledge resources flow inside multinationals, and their effective use and transfer have become difficult. It is crucial to have a sound communication infrastructure that forms an important foundation for the information transfer between the parent company and the overseas R&D branches and among overseas branches themselves. Presently, Chinese enterprises mainly use IT tools for communication in the process of internationalization of technological innovation, but on the whole the information technologies used are still relatively backward. Establishing a robust communication infrastructure is the first step for China's innovation internationalization.

3) Effectively improving the ability to absorb

More information sources and more channels for information transfer do not necessarily increase the transfer of technological information within an enterprise; the absorption ability of various branches of the enterprise is a key factor.

To enhance the absorption ability, enterprises first need to enhance the ability of their employees to recognize the value of new information and the ability to absorb and commercialize it. Organizations should acquire, train, and develop capable employees, especially those better than others in achieving deep exposure to external information sources, those who are technology gatekeepers good at screening and disseminating technological knowledge, and those who can understand both the field of information sources and that of information recipient and can help with knowledge transfer and dissemination between the two (Leonard-Barton, 2000).

For enhancing the absorption ability, to Chinese enterprises, the more important issue is to increase the driving forces of employees and work units alike to acquire knowledge. Presently, these enterprises do not have strong motivation to make use of information and knowledge resources in the process of internationalization of technological innovation. On the one hand, because new knowledge may disrupt the existing organizational practice and work routines and because knowledge transfer requires substantial input in practice and efforts, some teams and employees may not see related investment as worthwhile. On the other hand, the party to give out information may lose the sense of control brought about by the possession of special knowledge (Minvaeva et al., 2000). Without the motivation to acquire knowledge and information and to transfer as well as ultimately to commercialize them, absorption ability still cannot be improved. The organizations and teams in internationalized innovation activities should train their employees on values, policies, practices, systems, and structures to enhance their abilities and motivation for information exchange and should create an innovative culture conducive to the acquisition, development, and transfer of technological information resources. For example, enterprises can adopt performance-based reward system and internal promotion system as well as strengthen enterprise information sharing, make employees feel that they are trusted partners, and so on in order to promote the internal transfer of technological information and knowledge.

Globalization provides new knowledge distribution, but this does not mean that all countries and institutions are to profit equally. On the contrary, it seems that those that are already at the center of technological advancement will profit the most. This does not mean that developing countries are excluded. If the developing countries can adopt positive policies and principles to increase their access to knowledge and technology, they can also benefit from technology globalization (Archibugi & Pietrobelli, 2002).

5.5 Chapter summary

In this chapter, we started from the matching principle regarding the relationships among strategy, organization, and resource allocation. Based on the analysis of the present state of resource allocation in the globalized technological innovation system adopted by multinationals in the mature stage of internationalization of technological innovation, we studied the theories of resource allocation in the process of an enterprise's internationalization of technological innovation. Then, in combination with the analyses of the questionnaire survey results and case studies, we proposed several resource allocation models for different phases of internationalization of technological innovation by Chinese enterprises.

We believe that in the initial stage of internationalization of technological innovation by Chinese enterprises, innovation activities are mainly carried out domestically and there is principally one approach for resource allocation—the company's headquarters supplies human resources and capital resources to the overseas R&D branches; human resources and funds display a basic pattern of unidirectional flow from the company's headquarters to various branch units, but technological innovation information displays a different pattern of ongoing unidirectional flow from the branch units to the company's headquarters.

In the development stage of internationalization of technological innovation by Chinese enterprises, technological innovation activities display a pattern of coexistence of multiple branches in both overseas and domestic branches. The means for resource acquisition by the overseas R&D branches are no longer limited to the inputs mainly from the parent companies, the local recruitment and interunit personnel flow have greatly increased, and the sources of capital input have diversified—the company's headquarters, overseas subsidiaries, venture capital, and various noncommercial organizations can become investing parties. Information exchange has changed the flow from unidirectional between the layers of organizational hierarchy to bidirectional, thus leading to the existence of information exchange among the branch units.

The mature stage of internationalization of technological innovation is the time when enterprises truly carry out globalized technological innovation. The approaches for acquiring resources by the overseas R&D units have been diversified. Local recruitment of human resources and their interagency flow are the main characteristics; information sharing has been achieved, and the pattern of information flow is now multidirectional.

References

Archibugi, D. & Pietrobelli, C. The globalisation of technology and its implications for developing countries: windows of opportunity or further burden? Working paper series, 2002.

Cantwell, J. A. & Kosmopoulou, E. Determinants of internationalization of corporate technology (DRUID working papers No 01-08). Copenhagen: DRUID, 2001.

Chen, J. *Sustainable Development—New Perspective on Enterprise Technology Innovation*. Beijing: Science Press, 2001.

Cohen, W. & Levinthal, D. Innovation and learning: the two faces of R&D. *The Economic Journal*, 1989, 99(397): 569-596.

Dalton, D. H. & Serapio, M. G. *Globalizing Industrial Research and Development.* Washington, DC: U.S. Department of Commerce, Technology Administration Office of Technology Policy, 1999.

Davis, L. What location advantages do MNE research subsidiaries in Europe value highest? Rebild, Denmark: DRUID Summer Workshop, 2000.

Du, D. *Study on Location Model for R&D Globalization by Multinationals.* Shanghai: Fudan University Press, 2001.

Flaherthy, M. T. Coordinating international manufacturing and technology. In: Porter, M. (ed.). *Competition in Global Industries*. Boston: Harvard Business School Press, 1986: 83-110.

Gerybadze, A. & Reger, G. Globalization of R&D: recent changes in the management of innovation in transnational corporations. *Research Policy*, 1999, 28(2-3): 251-274.

Ghemawat, P. & Spence, A. M. Modeling global competition. In: Porter, M. (ed.). *Competition in Global Industries*. Boston: Harvard Business School Press, 1986: 61-79.

Golder, P. Insights from senior executive about innovation in international markets. *Journal of Product Innovation Management*, 2000, 17(5): 326-340.

Granstrand, O., Patel, P. & Pavvit, K. L. R. Multi-technology corporations: why they have “distributed” rather than “distinctive core” competencies. *California Management Review*, 1997, 39(4): 8-25.

Guellec, D. & Van Pottelsberghe de la Potterie, B. The internationalisation of technology analysed with patent data. *Research Policy*, 2001, 30(8): 1253-1266.

Gupta, A. K. & Govindarajan, V. Knowledge flows with MNCs. *Strategic Management Journal*, 2000(21): 473-496.

Hakanson, L. & Zander, U. International management of R&D: the Swedish experience. *R&D Management*, 1988, 18(3): 217-226.

Kogut, B. & Zander, U. Knowledge of the firm and the evolutionary theory of the multinational corporation. *Journal of International Business Studies*, 1993, 24(4): 625-645.

Kumar, N. Determinants of location of overseas R&D activity of multinational enterprises: the case of US and Japanese corporations. *Research Policy*, 2001, 30(1): 159-174.

Lane, P. & Lubatkin, M. Relative absorptive capacity and interorganizational learning. *Strategic Management Journal*, 1998, 19(5): 461-477.

Leonard-Barton, D. *Knowledge and Innovation*. Beijing: Xinhua Publishing House, 2000.

Lin, J. C. & Chai, Z. H. Analysis on the trend of internationalization of technological research and development in transnational corporations. *World Economy*, 1998(7): 35-39.

Minvaeva, D., Pedersen, T., Björkman, I., Fey, C. F. & Park, H. J. MNC knowledge transfer, subsidiary absorptive capacity and HRM. Working paper series, 2000.

Moenaert, R. K., Caeldries, F., Lievens, A. & Wauters, E. Communication flows in international product innovation teams. *Journal of Product Innovation Management*, 2000, 17(5): 360-377.

Niosi, J. & Godin, B. Canadian R&D abroad management practices. *Research Policy*, 1999, 28(2-3): 215-230.

OECD. Measuring globalisation: the role of multinationals in OECD economies. Paris: OECD, 1999.

Patel, R. D. & Pavitt, K. Large firms in the production of world's technology: an important case of non-globalization. *Journal of*

International Business Study, 1991, 22(1): 1-22.

Patel, P. & Pavitt, K. The wide (and increasing) spread of technological competencies in the world's largest firms: a challenge to conventional wisdom. In Chandler, A. D. et al. (ed.), *The Dynamic Firm: The Role of Technology, Strategy, Organization, and Regions*, 1998.

Rhinesmith, S. M. *A Manager's Guide to Globalization*. Chicago: Irwin Professional Publishing, 1996.

Tang L. Z. & Zhang S. Q. Trend analysis of R&D globalization by multinationals and China's countermeasures. *Asia-Pacific Economy*, 2001(3): 73-75.

Tolbert, A. S., McLean, G. N. & Myers, R. C. Creating the global learning organization (GLO). *International Journal of Intercultural Relations*, 2002, 26(4): 463-472.

Von Zedtwitz, M. Organizing global R&D: challenges, dilemmas and future trends. Miami: IAMOT2002, The 11th International Conference on Management of Technology, 2002.

Von Zedtwitz, M. & Gassmann, O. Market versus technology drive in R&D internationalization: four different patterns of managing research and development. *Research Policy*, 2002, 31(4): 569-588.

Wu, G. S. *Technology Innovation Management*. Beijing: Tsinghua University Press, 2000.

Zhao, C. P., Ling, D. C. & Liu, X. S. Study on motivation for R&D internationalization by multinationals. *Technoeconomics and Management Research*, 2001, 200(4): 84-85.

Chapter 6

Internationalization of Technological Innovation: Environmental Factors

Enterprise technological innovation is a set of multi-level and multi-link social activities; it is always done in a certain social environment and, therefore, is necessarily impacted by the social environment. Compared with the single home country-based technological innovation approach of the past, internationalization of technological innovation is to extend innovative activities to foreign countries, integrate innovative resources globally, and carry out knowledge and technology creation; as such, it is not only affected by the environment of the home country but also affected by the environment of the host country. Thus, the environment that internationalization of technological innovation faces now is even more complex, and the environmental factor has become an even more crucial factor affecting the outcome of internationalization of technological innovation.

The environmental factor not only affects decision-making of internationalization of technological innovation, but also impacts the process of internationalization of technological innovation by its effect on the operating mechanism of the process of internationalization. Conducting an in-depth analysis of the environmental factor at the enterprise level can help the enterprises in the following ways.

(1) Identify location advantages of various countries and make correct location selection for investment in internationalization of technological innovation, so as to adopt an appropriate strategy for innovation internationalization.

(2) Make use of different comparative advantages of different countries; make adjustment to the strategy of internationalization of technological innovation; consolidate the organizational structure, talent structure, capital structure, and so on; and dynamically update the internal operating mechanism of internationalization of technological innovation, so as to maintain and continuously improve efficiency.

Conducting an in-depth analysis of the environmental factor at the government level can help the government to take appropriate policy measures to create a better supportive environment for internationalization of technological innovation of Chinese enterprises.

6.1 Elements and characteristics of environmental factors for internationalization of technological innovation

6.1.1 Environmental factors for enterprises' internationalization of technological innovation

The environment of internationalization of technological innovation includes both the environment of the home country and the environment of the host country. By the nature of their different relationships to enterprise internationalization of technological innovation, environmental factors can be separated into four levels (Figure 6-1): national social cultural level, national policy level, national resource level, and national industry level.

Environmental factors at the national social cultural level mainly include a country's languages, national value orientation, religious beliefs, esthetic ideas, customs, etc.; the factors at the national policy level include international technology transfer policy, policy of S&T, foreign trade policy, intellectual property protection policy, etc.; the factors at the national resource level include the size of the country's market, its human resources in technological innovation, its environment for technological innovation, its technological infrastructure, etc.; the factors at the national industry level include enterprise competitive status, user and supplier relationships, etc.

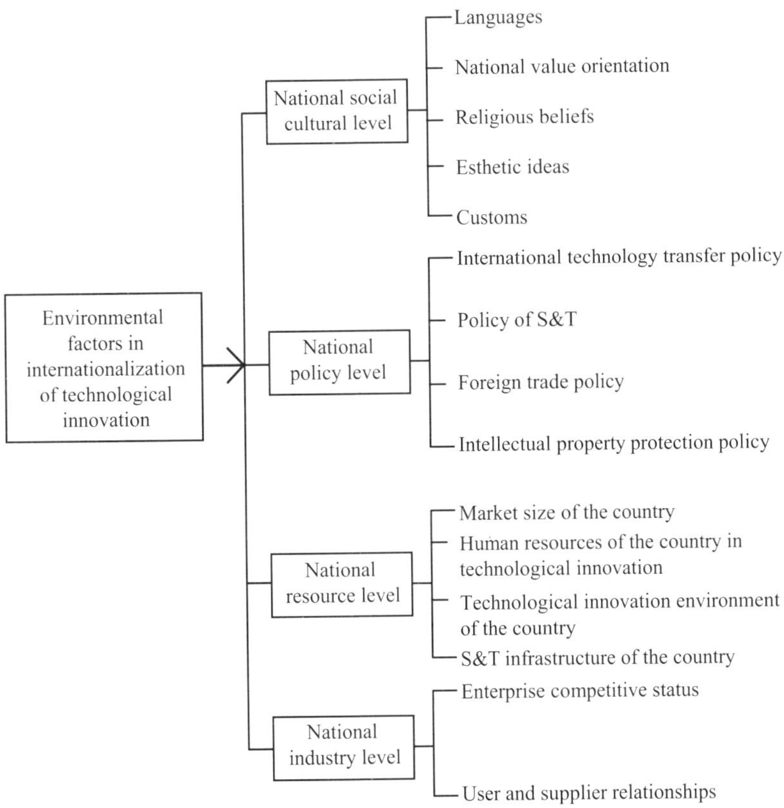

Figure 6-1 Environmental factors in internationalization of technological innovation

6.1.2 Environmental characteristics of enterprises' internationalization of technological innovation

Compared with traditional technological innovation, the environment faced by enterprises' internationalization of technological innovation has two notable characteristics: diversity and complexity.

First, the environment faced by internationalization of technological innovation is diversified.

This diversity is reflected in the differences among different countries in social environment, cultural environment, governmental policies, market environment, and resource environment for technological innovation. When an enterprise carries out internationalization of technological innovation activities in different countries, such activities will be affected by the environmental factors of all of these countries. Thus, there are more environmental factors and more ways for these factors to have an impact compared to those of the traditional technological innovation.

Second, the environment faced by internationalization of technological innovation is complex.

For a multinational at a high level of internationalization of technological innovation, its innovation internationalization has multiple strategic objectives, takes on many forms, employs multiple overseas technological innovation units, and is carried out in multiple host countries, the different environmental factors of different host countries work on the enterprise's internationalization of technological innovation system and the impact of the environment on the enterprise's internationalization of technological innovation is thus more profound and more complex.

The complexity of the environmental factors is mainly reflected in the differences and conflicts among the cultures of the host countries, the differences and conflicts among the government policies of the host countries, the diversity and variety of the market conditions, and the diversity and variety of the innovative resources that work on the internationalized technological innovation system of the business enterprises. The uncertainty of the environmental factors increases and forces the enterprises to make hard choices when making decisions about their technological innovation; as a result, enterprise innovation management is put in a situation teemed with numerous conflicts and contradictions.

6.2 Impact of host country's environmental factors on enterprises' internationalization of technological innovation

Whether or not an enterprise should carry out the process of internationalization of technological innovation and how to select the location for investment in the process are the first issues to consider in the decision-making of internationalization of technological innovation. Analysis of environmental factors is the basis for decision-making; satisfactory environmental factors are the primary condition of location selection for investment in enterprise internationalization of technological innovation.

6.2.1 Impact of host country's environmental factors on internationalization of technological innovation that can result in location advantages or location disadvantages

J. Dunning (1988), a famous expert on multinationals,

in his research on international direct investment, proposed a comprehensive eclectic theory of international production. At the core of the eclectic theory are three advantage theories: ownership-specific advantage theory, internalization-specific advantage theory, and location advantage theory. According to Dunning, ownership-specific advantages—also called monopoly advantages—refer to all the tangible and intangible comprehensive advantages of a business that enterprises of other countries do not have or will have difficulty acquiring, such as assets, scale, technology, markets, etc.; internalization-specific advantages refer to the advantages that come from low trade costs and high transaction efficiency by forming the internal transaction system of an international company; location advantages refer to the advantages of the host country's environmental resources that attract international investment. Ownership-specific advantages and internalization-specific advantages determine the ability to make direct investment overseas, but they do not explain the overseas direct investment actually made by an enterprise; on the premise that an enterprise has the first two advantages, the sufficient conditions of overseas direct investment are the location advantages of the host countries (see Figure 6-2).

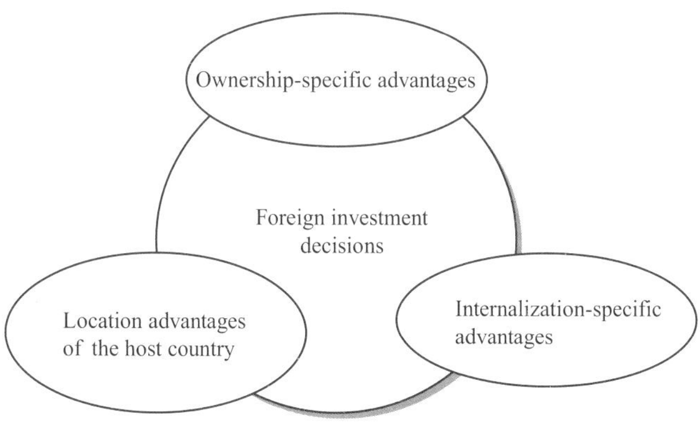

Figure 6-2 Dunning's theoretical model for international investment decisions

Granstrand (1992)—a renowned Swedish expert on internationalization of technological innovation—believed that the determinants of enterprises' internationalization of technological innovation include the internal promoting force, the internal hindering force, and the environmental factor (see Figure 6-3). At the corporate level, the centrifugal force of internationalization, that is, the promoting force, includes two aspects—the force of demand orientation and the force of supply orientation. The force of demand orientation includes technology transfer to subsidiaries, integration of local production, subsidiary regional expansion, government control, and proximity to local customers and market. The force of supply orientation includes overseas M&A, proximity to overseas S&T, proximity to low-cost/low-efficiency R&D human resources, and so on. The centripetal force of internationalization of technological innovation is the hindering force that includes the need to control R&D, concerns about leaking trade secrets, the need to be close to domestic market, economy of scale for R&D, cost of collaboration and exchange, historical inertia, etc. Environmental factors include these factors at the national and international levels. At the national level, the decision-making factors for internationalization of technological innovation mainly include the culture of the enterprise's home country and the host country, the structure of the governing system, government policy, internal/external input/output market condition, and so on. At the international level, the decision-making factors include intercountry policies, agreement and system, integration of trade barriers, internationalism, global environmental factors, and so on.

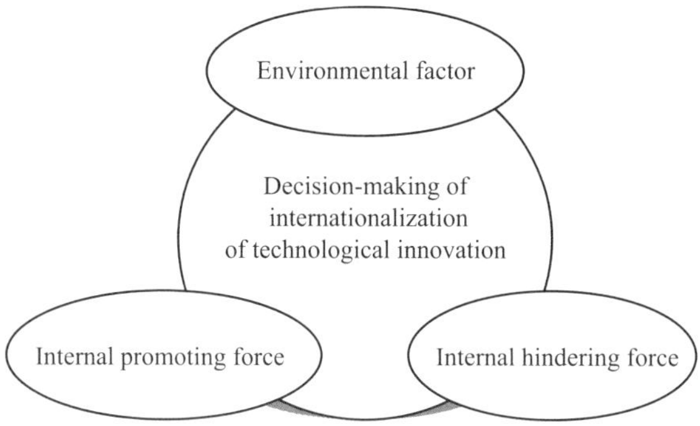

Figure 6-3 Granstrand's determinant model of internationalization of technological innovation

In Dunning's eclectic theory of international production, the corporate ownership-specific advantage theory and internalization-specific advantage theory are not targeted at enterprises' internationalization of technological innovation. However, no matter whether it is for international direct investment or for investment in internationalization of technological innovation, all the environmental resource advantages of the host country form the sufficient condition for investment. Granstrand's determinant model of internationalization of technological innovation comprehensively analyzes the various determining factors of enterprises' internationalization of technological innovation; the internal and environmental factors of an enterprise work together to determine whether or not the enterprise will implement the strategy for internationalization of technological innovation.

The author believes that investment in internationalization of technological innovation is determined jointly by the driving force and the attracting force for investment. The driving force is the internal drive of an enterprise to invest in internationalization of technological innovation. Such driving forces include demand-oriented force and supply-oriented force and exist in the investing enterprise itself. The attracting force refers to the influence of the host country on internationalization of technological innovation, which comes from the environmental factors of the host country. In a decision by an enterprise to invest in internationalization of technological innovation, the internal driving force and the attracting force (which is the result of the environmental location advantages of the host country) influence each other and are indispensable (Figure 6-4). When an enterprise has the internal driving force for internationalization of technological innovation, it only has the necessary condition for such internationalization. When the environmental factors of the host country are supportive of internationalization of technological innovation, they become the location advantages of internationalization of technological innovation and such advantages will influence the enterprise's internal driving force to bring about investment for internationalization of technological innovation in the host country.

Internal driving force for investment + Attracting force of host country's environmental factors = Investment for internationalization of technological innovation

Figure 6-4 Impact of environmental factors on investment for internationalization of technological innovation

6.2.2 Mechanism by which environmental factors of host country affect enterprises' internationalization of technological innovation

(1) Factors at the national social cultural level

A country's social environment generally includes its demographics, income distribution, family composition, and so on. A country's cultural environment refers to the commonly held ideas, emotions, and combined behaviors, and it includes language, education, value system, religious beliefs, esthetic ideas, customs, etc.

The effect of the factors at a country's social cultural level has two aspects.

First, language barriers and cultural differences affect the exchange and communication between an innovative organization and its environment and within the innovative organization itself, thereby affecting the efficiency of internationalization of technological innovation and the location choice.

A common vision and the goal orientation form the basis for effective operation and success of a technological innovation organization. In an internationalized technological innovation organization, employees have different educational backgrounds, value systems, religious beliefs, esthetic ideas, and customs, and thus, it is difficult for the organization to develop a unified vision.

The primary task of technological innovation is to search and access innovative information. Language barriers and cultural differences of the host country will have a negative effect on information access to international innovation activities.

The efficiency and outcome of a technological innovation project depend on the level of exchange and communication within the organization. Knowledge codification is the key factor for the successful exchange and communication within an internationalized technological innovation organization. New product development in an international environment may stop the knowledge codification process. This is mainly because of the following reasons: (i) different units and functional departments of an organization will develop their own special "languages," subcultures, and methods for exploring and analyzing the environment; (ii) the linguistic, cultural, and other environmental differences among the countries make encoding more difficult (Moenaert et al., 2000).

Such barriers in exchange and communication that are caused by the differences in cultural environment will reduce the efficiency of internationalization of technological innovation.

It is because of such considerations and concerns that when an enterprise is making the location choice for internationalization of technological innovation, the first considerations are usually host countries that have very similar languages and cultures (Bartlett & Ghoshal, 2002).

Second, the specific social and cultural environment of a country makes the local market conditions and consumer preferences different. Such difference, in turn, will affect the process of internationalization of technological innovation.

The ultimate goal of technological innovation is to bring the fruits of R&D to the market. Thus, in the process of technological innovation, from information access and idea formation to research, production, and marketing, every phase must be accomplished by considering the specific requirements of the local market conditions and customer needs. The unique social environment and cultural environment of a specific host country affect the local market conditions and consumer preferences and ultimately affect the process of internationalization of technological innovation and its outcome.

To overcome the exchange and communication barriers that arise out of the differences in cultural environment and to adapt to the effects of the local social environment and cultural environment on technological innovation, an enterprise must make adjustments to its organizational structure and resource allocation as well as management (Figure 6-5). For an effective organizational structure, it is a must to adopt a parallel structure, set up an international steering committee, and let it operate concurrently with the hierarchical structure; for configuring human resources, it is a must to select team leaders who can bridge cultural differences, control information flow, and reconcile the needs of different people and localize R&D personnel—both are very necessary; and for management, the key factor is to strengthen the cross-cultural coordination and integration.

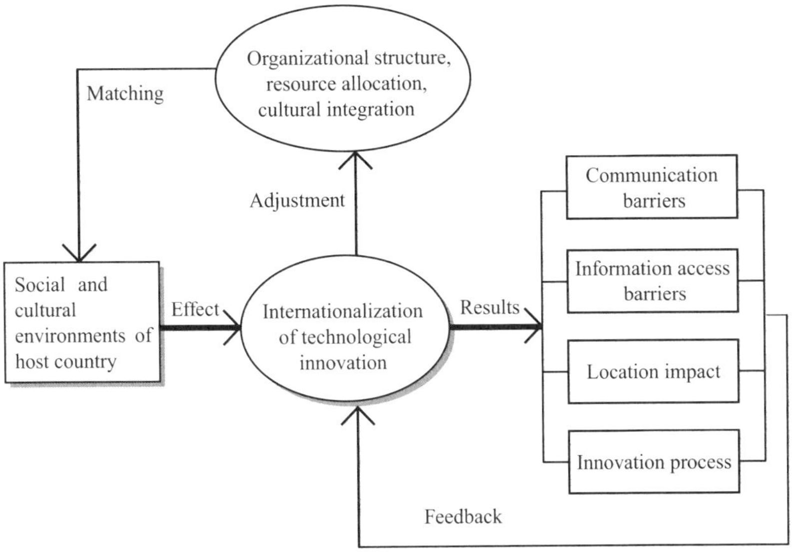

Figure 6-5 Interactive mechanism of social and cultural environments of host country and enterprises' internationalization of technological innovation

(2) Factors at the national policy level

The policies of different host countries related to enterprises' internationalization of technological innovation can be divided into those that are supportive and those that are restrictive of investment from foreign countries for internationalization of technological innovation. Incentives given by a host country for foreign investment in internationalization of technological innovation include the following: tax reduction or exemption for such foreign investment, strengthening the protection of intellectual property rights, facilitating patent application by foreign companies for their R&D activities in the host country, improving S&T infrastructure, and adding investment in technological innovation as an additional condition to direct foreign investment. Restrictive policies of a host country mainly target R&D activities of the information gathering type and R&D M&As.

In fact, on the one hand, the spillover effect of technological innovation in the host country by a foreign company helps improve the technological innovation capabilities of the host country. Therefore, many countries formulate policy incentives to attract overseas R&D investment, in an effort to strengthen their own S&T and technological innovation capabilities. For example, on policy formulation related to technological innovation, OECD countries have different policy measures, but they have the same four policy objectives and theoretically all these policy objectives try:

- to prevent domestic R&D from going abroad;
- to encourage foreign companies to invest in its R&D;
- to maximize the benefits from foreign R&D investment; and
- to deepen the links between domestic and global technological innovation systems and to benefit more from the global innovation system. (China Society of Science and Technology Indicators, 2002)

On the other hand, international technological innovation activities of the technology search and monitoring type can result in leakage of the host country's technological achievements; R&D activities through M&As by a foreign country can also weaken the technological innovation capabilities of the host country. Therefore, the host country may also restrict foreign R&D activities in some specific areas and forms. Especially, in the high-tech field, some OECD countries have implemented various policies stipulating strict conditions to be met before foreign countries can participate in the national research bases (OECD, 1999).

Because different host countries have different national policies and have different rules about country-specific cross-border information inflow/outflow related to international technological innovation projects, uncertainties in project management of internationalization of technological innovation have increased.

(3) Factors at the national resource level

A host country's national resource is a generalized concept, and factors at the national resource level of the host country refer to the environmental factors related to technological innovation, which are parts of the environment for technological innovation and constitute the resources in relation to internationalization of technological innovation (see Figure 6-6).

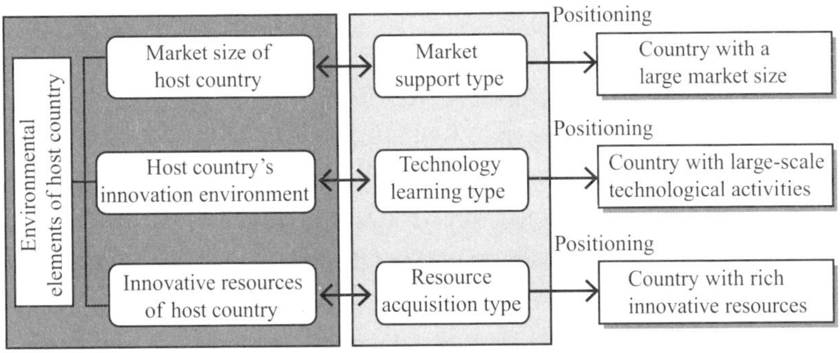

Figure 6-6 Impact of host country's environmental factors on location choice for internationalization of technological innovation

First, factors at the national resource level of the host country greatly affect the location selection of internationalization of technological innovation (Kumar, 2001).

Market size of the host country is the primary factor influencing location choice for internationalized technological innovation activities of the technology support type. Because of the economy of scale related to innovative activities, it is inefficient to adapt to a very small-scale market. Thus, the higher the degree of market penetration of the enterprise in the host country, the larger the market size of the host country; the greater are the chances of overseas R&D activities to enter the host country.

Human resources for technological innovation and the accessibility of lower cost or more advanced scientific infrastructure in the host country have a great deal of influence on the location selection for internationalization of technological innovation of the resource acquisition type. Rich well-trained R&D human resources and the high accessibility of lower cost and more advanced scientific infrastructure make a country more attractive to foreign R&D activities; a country with rich and low-cost R&D human resources will induce multinationals to locate part of their R&D activities in the country in order to reduce their global R&D costs (Kumar, 2001).

A host country's technological innovation environment has a direct appeal to the internationalization activities of the technology learning and technology searching types. Countries with large-scale technological activities and world-class technology centers of excellence have more opportunities to attract foreign R&D investment that is aimed at benefiting from the spillovers of R&D of the host countries.

Second, resource environment of the host country increases the idea sources of internationalization of technological innovation, expands the space for technological innovation, and enriches the technological innovation resources, and it also makes new demands for the management of internationalization of technological innovation (Figure 6-7).

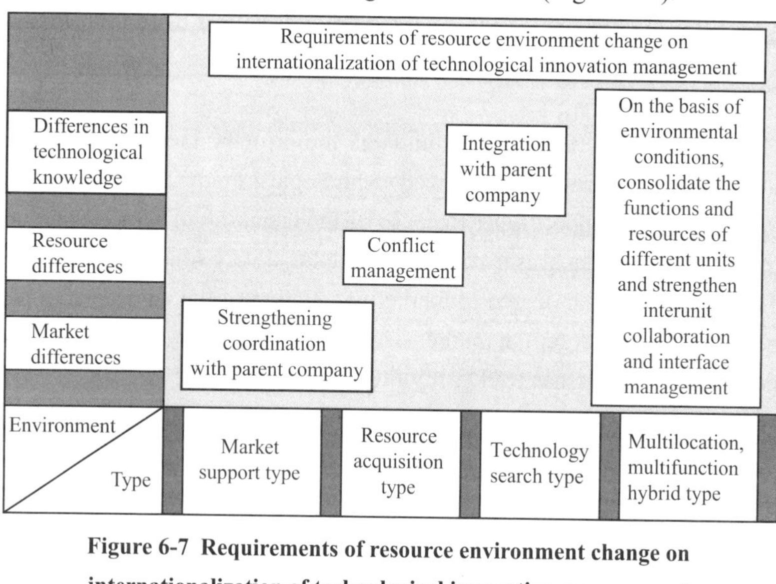

Figure 6-7 Requirements of resource environment change on internationalization of technological innovation management

Market-support-type internationalization of technological innovation must be closely connected to the local market conditions. Due to the differences in market conditions and consumer preferences, organically combining local market conditions with existing technology of the company becomes significant; the organic coordination of local technological innovation unit and parent company's R&D center in organization, resources, communication, and so on is the primary factor for successful local innovation.

When internationalization of technological innovation of the resource utilization type is to obtain the rich resources of the host country, resource integration is the key. Local R&D staff members have a cultural background and value system that are different from those of the parent company personnel, and thus, there should be more communication and coordination in the local innovation unit; managing cultural conflicts is essential for maintaining organizational efficiency.

The key point in internationalization of technological innovation of the technology search and technology learning types is how to acquire overseas technology and knowledge and transform them into the company's technological capabilities; close connection in the innovative process between parent company and innovation unit, organic integration of the managerial function, and the implementation of an integration strategy are very important.

Meanwhile, for an enterprise that is carrying out global technological innovation activities in many different locations, it is important to consolidate the functions and resources of the various geographically dispersed innovation units faced with different environmental conditions, to reconstruct the organizational structure and processes, and to strengthen the collaboration and interface management among its various units in order to maintain the efficiency of its innovation.

(4) Factors at the industry environment level

1) Factors at the host country's industry environment level affecting the location choice and mode of foreign investment in the host country's internationalization of technological innovation

First of all, the degree of technological competition in a certain industry of the host country will affect the location choice and the mode of foreign investment in the host country's internationalization of technological innovation. If local enterprises are too competitive in a particular industry of the host country, investment by foreign enterprises in this industry of the host country will be reduced. Meanwhile, enterprises in other industries will be drawn to invest (Cantwell & Kosmopoulou, 2001). In other words, when local enterprises of the host country become too competitive in a particular industry, foreign companies will not use this country as its main base for technological innovation; it will be very difficult for internationalized technological innovation investment to enter either for the technology support type or for locally based knowledge creation, but there will be more internationalized technological innovation investment in the technology monitoring type for the purpose of monitoring the technological trend of the competitors or in the technology learning type for the purpose of learning related technology.

Second, when the technological capabilities of a foreign enterprise are determined by the interactive innovation of the enterprise and the suppliers as well as consuming parties of the host country and by the close links among the various industries, the enterprise will make large investments in technological innovation in the host country (Cantwell & Kosmopoulou, 2001).

2) Factors at the host country's industry environment level affecting the process and management of technological innovation in the host country by foreign enterprises

When local enterprises of the host country are very competitive in a particular industry, technological innovation by foreign enterprises in the host country will be mainly based on technology monitoring and technology search. Because the host country is not a main technological innovation base of foreign enterprises, they will allocate relatively limited technological innovation resources in the host country and will carry out technological information gathering in the host country, but their technological innovation processes will be mainly done in respective headquarters or innovation bases. Links with the host country are not the focus of technological innovation management, but close links and coordination with the enterprise's technological innovation network are crucial to the success of innovative activities.

When the technological capabilities of a foreign enterprise are determined by the interactive innovation of the enterprise and the suppliers as well as by the consuming parties of the host country and the close links among the various industries, the enterprise will make the host country its technological innovation base or set up a technological innovation center. In the process of technological innovation by an enterprise, information gathering, idea formation and development, and marketing are highly dependent on the environmental conditions of the host country and coordination with the host country becomes an important factor in innovation management.

6.3 Impact of home country's environmental factors on enterprises' internationalization of technological innovation

International environment affects enterprise internationalization of technological innovation, and the environmental factors of the home country greatly affect the decision-making related to enterprise internationalization of technological innovation. For an enterprise, policies, resource environment, and industry organization of the home country are the factors directly affecting its decision-making about internationalization of technological innovation.

6.3.1 Mechanism by which policies of home country affect enterprises' internationalization of technological innovation

(1) Effect of home country policies on enterprise internationalization of technological

innovation determined by home country's technology development objectives

The technological innovation policy formulated by a country is linked with the country's understanding of technological innovation. As observed in the OECD findings earlier, technological innovation capabilities contribute positively to a country's economic growth, productivity growth, employment, and trade performance (China Society of Science and Technology Indicators, 2002) and, at the same time, have positive spillover effects on other sectors of the economy. Therefore, the policy objectives of OECD countries converge into one, which is, through policy influence as much as possible, to make the home country benefit from foreign R&D investment in the home country and from overseas R&D investment by the home country, so as to enhance home country's technological prowess.

National policies related to internationalization of technological innovation by the home country's enterprises can also be divided into those that are supportive and restrictive. When a country is committed to improving its technological capabilities, it will adopt policies encouraging its enterprises to engage in technological innovation and attracting foreign investment in its R&D such as using public spending to fund enterprise technological innovation, tax cuts, or exemptions for enterprise R&D. These policies will on the one hand increase the resources for enterprise technological innovation and on the other hand reduce the cost of enterprise technological innovation; thus, to domestic enterprises, such policies will decrease their motivation to carry out internationalization of technological innovation, and for foreign enterprises, such policies will make the country more attractive to their investment in technological innovation.

The construction of a national innovation system aimed at encouraging technological innovation optimizes the innovative environment of the home country; such efforts encourage domestic enterprises to carry out major innovative activities in the home country and at the same time make the country more attractive to foreign R&D investment.

The technology transfer policy of a country is also a factor in determining the tendency of its enterprises to carry out internationalization of technological innovation. When a country wants to prevent its domestic enterprises from leaking technology secrets while making overseas R&D investment, it adopts technology transfer policies that will hinder investment in foreign R&D by its enterprises.

Enterprises' internationalization of technological innovation is also positively correlated with national trade policies. When a country adopts an active foreign trade policy, its enterprises—due to their expanding business operations—will have stronger tendency and needs to carry out internationalization of technological innovation (Cantwell, 1995).

(2) Home country's policy formulation related to enterprises' internationalization of technological innovation affected by the nature of such internationalization

Assuming that a country is committed to protecting and enhancing its technological capabilities, the country's policy formulation will be affected by the nature of internationalization of technological innovation of its enterprises. For R&D investment purely for adapting to overseas market conditions, since the impact on domestic economy is limited, there will not be much attention to it at the national policy level; in case of R&D investment abroad for the purpose of innovation, since it brings about a "new technology division of labor" between the parent company and its overseas subsidiaries, such investment is hugely beneficial to both the home country and the host country, and thus, the home country policy will be supportive of it; and for R&D activities aimed at searching for knowledge abroad, since they eventually will transfer tacit knowledge that is basically "localized" and thus will improve the efficiency of domestic production and technological innovation, the home country policy will also be supportive (China Society of Science and Technology Indicators, 2002).

(3) A country's policy toward internationalization of technological innovation by domestic enterprises being related to its technology development (China Society of Science and Technology Indicators, 2002)

For the smaller among the Nordic countries, the degrees of internationalization of the production and operations of their enterprises are higher than those of OECD countries on average, but they lack technological innovation resources because of their small size, and thus, their national policies tend to encourage enterprises' internationalization of technological innovation, helping their enterprises to find appropriate places in the basic structure of global research in order to improve their innovative capabilities. For other countries like Germany and Sweden, they have good technological foundations, but because they are committed to ensuring their attractiveness to foreign industrial research in the process of economic globalization and at the same time preventing domestic enterprises from leaking research results, their national policies do not lean toward supporting enterprises' internationalization of technological innovation. For some countries, such as Poland, Hungary, and R.O. Korea, of the catching-up type in technology development, they are committed to obtaining new technology and enhancing their technological innovation capabilities through internationalization of technological innovation, commodity trade, and the return of their overseas scientists, and thus, their national policies are notably supportive of internationalization of technological innovation.

6.3.2 Mechanism by which innovative resource environment of home country affects enterprises' internationalization of technological innovation

A country's technological innovation resources or the lack thereof either drives or hinders internationalization of technological innovation of its enterprises (Figure 6-8).

Level of home country resource environment	Low	High
High	Main technological innovation activities concentrated in the home country: technology search and technology learning	Core technologies concentrated in the home country, supplementary technologies internationally innovated
Low	Not sure	High degree of internationalization of technological innovation; resource utilization type, technology learning type, and market support type

Degree of internationalization of enterprise operations

Figure 6-8 Relationship between home country resource environment and enterprises' internationalization of technological innovation

Relative scarcity of technological innovation resources in a country is one of the driving forces of internationalization of technological innovation by its enterprises. Traditional assumption is that large enterprises in small industrialized countries, when formulating technological strategies, show a strong tendency to internationalize and that the tendency to internationalize is not so obvious for multinationals from large countries. Because of the geographic restrictions of small countries, their large enterprises are forced to internationalize faster (Cantwell & Kosmopoulou, 2001). This assumption makes sense if all other factors remain unchanged. Because of the small market size of small countries, large enterprises are pushed to develop internationally and enterprises' internationalization of technological innovation is directly linked with the degree of its international production and operations (Cantwell, 1995). For example, the probability of a Japanese multinational to carry out overseas R&D goes up with its size, the scale of its overseas operations or production, and the intensity of its technology (Kumar, 2001). Talent for technological innovation in a small country are relatively scarce, the technological innovation environment cannot support the needs of large-scale technological innovation of large enterprises, and thus large enterprises are pushed to look for innovative resources worldwide. In this regard, the enterprises of small European countries are excellent examples: the small size of their home countries is the reason for the geographic diversification of their R&D activities (Kumar, 2001).

For enterprises of countries with a large market size and rich technological resources, since the technological resource environment of the home country is very supportive of enterprise technological innovation or since the market of the enterprises is mostly domestic, their technological innovation activities are mainly concentrated in the home country. Multinationals with high levels of internationalization in both technology and production as well as operations still tend to base the core R&D activities—which produce its core technology—within their home countries (Kumar, 2001).

However, because the decision-making regarding internationalization of technological innovation is affected by multiple factors, technological innovation resource environment of the home country alone cannot explain whether the technological innovation of an enterprise will stay inward or go outward.

6.3.3 Mechanism by which industry organization environment of home country affects enterprises' internationalization of technological innovation

The intensity of technological competition in an industry affects the degree and mode of an enterprise's internationalization of technological innovation (Cantwell, 2001). If enterprises in a particular industry of a country are too competitive, they will be pushed to make overseas investment, but investment by foreign enterprises in that industry of its own country will be reduced (Figure 6-9).

		From home country viewpoint	
		Investment from foreign countries	Investment in foreign countries
Analysis dimension	Industry	High degree of internationalization (H)	Regional barriers in industry (L)
	Technology	Exploration of technology diversification (M)	Absorption of technology of industry, enterprise local diversification (M)

Figure 6-9 Effect of dominant domestic enterprises on degree and mode of internationalization of technological innovation

Note:

H: High degree of internationalization in industry or technological field;

M: Medium degree of internationalization in industry or technological field;

L: Low degree of internationalization in industry or technological field.

First of all, in industries where local enterprises are the technology leaders, the degree of internationalization of technological innovation by the local enterprises will be relatively higher (Cantwell, 1989). However, such multinationals make use of internationalization of technological innovation network mostly for increasing its degree of relative diversification (Cantwell & Piscitello, 2000), but in the main technology field of its own main business, they will still lean toward domestic development.

Second, from the viewpoint of inward investment, when powerful local enterprises are making domestic investment, they will always invest in their core technology field, and such practice will inhibit substantive progress in the local industry by foreign enterprises. Meanwhile, powerful foreign enterprises in other industries may be interested in this technology center of excellence and thus may decide to develop related technologies of their own near the center so as to achieve their own industry diversification. Since their main businesses are in other fields, they do not constitute direct competition to the leading local enterprises.

When enterprise technological innovation is dependent on its close relationship with either local downstream customer businesses or upstream suppliers, its level of internalization will be comparatively low.

When an enterprise uses local downstream customer companies or upstream suppliers as the source of innovation and source of important feedback for internal innovation and sees them as the foundation for enhancing technological competitiveness, since such relationship emphasizes mutual trust and the creation of localized specific knowledge, there should be commitment of substantive resources, "assets that have continuity and cumulativeness in the time dimension," and thus, the local industry group will tend to maintain the current status in location choice for innovation (Lundvall, 1985, 1988). The more the local industry group for the local market is dependent on the local suppliers to provide special creative devices or other resources, the less the group will be willing to internationalize when formulating R&D strategies.

When the technical capabilities of a local enterprise depend on the interactive innovation between the manufacturing business and the consuming business as well as on the close ties among various industries, we will probably see the situation shown in Figure 6-10. First, the level of overseas investment in innovation by the enterprise will be low, which will be even lower than the level of local investment by foreign enterprises. In this situation, the level of inward and outward international investments in the industry will be below that of the corresponding technological fields (because enterprises in the related industries can probably enter the field abroad easily in order to carry out diversification or to gain proximity to local center of excellence; local enterprises in this industry will find it very difficult to establish similar relationships or to operate independently in other locations and foreign enterprises will find the local relationship network a barrier of entry in the industry).

		From home country viewpoint	
		Investment on innovation in foreign countries	Investment in foreign countries
Analysis dimension	Industry	Dependency on local customers or suppliers (L)	Barriers of entry due to close ties among local enterprises (L)
	Technology	Exploring customer or supplier capacity (M)	Attracting foreign enterprises of related industries to develop diversified technology locally (M)

Figure 6-10 Effect on internationalization of technological innovation when sources of innovation are mainly local supply and sales network

Note:

M: Medium degree of internationalization in industry or technological field;

L: Low degree of internationalization in industry or technological field.

6.4 Analysis of environmental factors and policies of internationalization of technological innovation for Chinese enterprises

The results of our study show that Chinese enterprises began to carry out internationalization of technological innovation not long ago, most enterprises set up R&D alliances with multinationals or technologically advanced companies in order to gain technology transfers, few enterprises have truly gone abroad to set up R&D institutions, existing overseas R&D institutions are mainly engaged in technology search and monitoring activities, and the proportion of overseas R&D institutions of the market support type is small. It can be said that R&D internationalization by Chinese enterprises is still in the initial stage. Investigating and analyzing the environmental factors of internationalization of technological innovation by Chinese enterprises have important policy significance to enterprise decision-making in internationalization of technological innovation and to related government guidance at the macro level.

6.4.1 Analysis of the effect of host country's environmental factors on internationalization of technological innovation for Chinese enterprises

(1) Host country's technological innovation resources constitute the main factor of attention for Chinese enterprises when carrying out internationalization of technological innovation

Considering the motivation of Chinese enterprises when carrying out internationalization of technological innovation (Table 6-1), statistical results of the survey showed that gaining proximity to overseas R&D environment as well as absorbing spillover of R&D results, using good overseas R&D infrastructure, recruiting outstanding R&D personnel from abroad, carrying out cooperative R&D projects overseas, and learning from technology transfers to China have received very high average scores and very few enterprises believe that they are not important.

Table 6-1 Motivation evaluation of R&D internationalization by Chinese enterprises

Factor	Average score	Most important	Important	Medium	Not important	Least important
1. Provide necessary technology support in product and process for local subsidiaries	2.64	14.3%	21.4%	14.3%	28.6%	21.4%
2. Introduce new products to the local market as soon as possible	2.71	14.4%	7.1%	21.4%	35.7%	14.4%
3. Adapt to the special needs of consumers in different parts of the world and produce specialized products needed	2.57	14.3%	0	35.7%	28.6%	21.4%
4. Get close to overseas R&D environment and absorb spillover of the R&D results	4.21	31.6%	57.9%	10.5%	0	0
5. Make use of good R&D infrastructure overseas	3.14	7.1%	21.5%	28.6%	0	42.8%
6. Recruit outstanding R&D personnel from abroad	3.0	14.3%	28.6%	21.4%	14.3%	21.4%
7. Carry out cooperative R&D projects overseas and learn from foreign technology transfers to China	4.48	57.1%	33.3%	9.5%	0	0

Particularly for two indicators, namely "carry out cooperative R&D projects overseas and learn from technology transfers to China" and "get close to overseas R&D environment and absorb spillover of the R&D results," most enterprises believe that they are important or very important (90.4% and 89.5%, respectively). This suggests that the objective of making use of the S&T and R&D resources of different countries of the world to carry out technology search, technology learning, and technological innovation so as to more effectively enhance enterprise technological strength is the main motivation of R&D internationalization by Chinese enterprises. The host country's technological innovation resources constitute an important factor of attention for Chinese enterprises in the process of internationalization of technological innovation.

For the regional distribution of internationalization of technological innovation by Chinese enterprises, according to the survey, 40% of the overseas R&D branches established by Chinese enterprises are concentrated in the United States, 13.3% in Europe, and 0.7% in Japan. Among the technology alliances established with foreign enterprises, those established with multinationals of developed countries, such as the United States, European countries, and Japan, account for 93.4% and those established with companies of developing countries account for 6.6%. Enterprise activities for internationalization of technological innovation are mainly focused on the world's technology centers in the developed countries, such as those in Europe and the United States. This also shows that the accessibility of a host country's technological innovation resources has important impact on the technological innovation activities of Chinese enterprises.

The results of our survey and interviews showed that Chinese enterprises in different industries, including high-tech (such as IT), home appliance, and machinery manufacturing (such as mobile) industries, have different focuses of attention regarding a host country's environmental resource conditions in the process of internationalization of technological innovation (Figure 6-11).

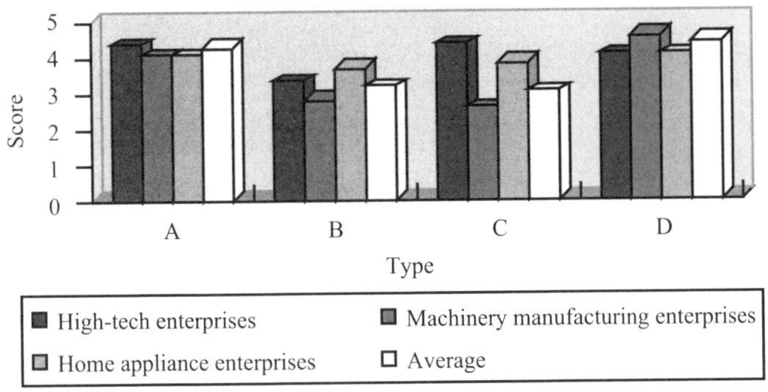

A: Get close to overseas R&D environment and absorb spillover of the R&D results;

B: Make use of good R&D infrastructure overseas;

C: Recruit outstanding R&D personnel from abroad;

D: Carry out cooperative R&D projects overseas and learn from foreign technology transfers to China.

Figure 6-11 Attention by various industries to innovative resources in host country

Enterprises in the IT industry pay most attention to the host country's innovative environment and technology talent. For example, Huawei has set up overseas R&D institutions in the United States (Dallas and Silicon Valley), Sweden, India, and Russia. The main purposes of the institutes in the United States and Sweden are to get involved in the local innovative environment; to monitor the development in mobile telecommunications technologies, such as optical network, GSM, and WCDMA; and to absorb the spillover of local technological innovation. The main purposes of Huawei's institute in India are to learn from India's software development process and its method for quality control and to make use of India's software research resources. Similarly, ZTE, for the same purposes, has set up overseas R&D institutions in the United States (Silicon Valley, New Jersey, and San Diego) and R.O. Korea, and Lenovo has also set up a joint laboratory with an American company in the United States (Silicon Valley).

Enterprises in the home appliance industry pay most attention to the host country's innovative environment and technology cooperation. Shandong Xiaoya Group in Jinan has first set up a joint laboratory with Toshiba of Japan for jointly developing its first and second generation washing machines, and at the same time, after comparing the accessibilities of Japan's and Europe's technological resources, in order to access Europe's environment and resources for advanced home appliance R&D, it has set up its new generation product development laboratory in Germany.

Enterprises in the machinery manufacturing industry represented by the automobile manufacturing sector pay most attention to cooperative learning. International innovative activities in this field by Chinese enterprises are basically carried out with the automotive manufacturers of developed countries, such as the European countries and the United States, with the most typical being FAW-Volkswagen, Shanghai Volkswagen, DFAC Shenlong, Guangzhou Honda, etc. Without any exception, these companies first consider the accessibility of the host country's technological innovation resources and the quality of the technological innovation environment.

(2) Host country's market condition not being a major environmental factor for internationalization of technological innovation for Chinese enterprises

As listed in Table 6-1, indicators—including providing necessary technology support in product and process for local subsidiaries, introducing new products to the local market as soon as possible, adapting to the special needs of consumers in different parts of the world, and produce specialized products—have received low average scores. Most enterprises believe that their impact on internationalization of technological innovation is either ordinary or not important. This shows that Chinese enterprises carry out limited market-support-type activities for internationalization of technological innovation for the market needs and that host country's market condition is not a major environmental factor for internationalization of technological innovation by Chinese enterprises.

At the industry level, high-tech enterprises pay more attention to host country's market condition. This is related to the marketing efforts in specific international markets by several major telecommunications equipment manufacturers, such as ZTE's marketing effort in Southeast Asia, West Asia, North America, Eastern Europe, Latin America, Africa, and so on and Huawei's marketing effort in Germany, Russia, Spain, Singapore, Thailand, R.O. Korea, Brazil, Kenya, and so on. Enterprises in the automotive industry pay least attention to host country's market condition because their sales are mainly concentrated in the domestic market (see Figure 6-12).

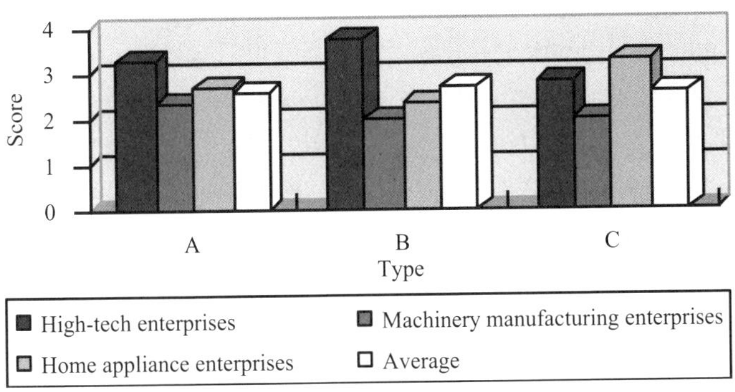

A: Provide necessary technology support in product and process for local subsidiaries;

B: Introduce new products to the local market as soon as possible;

C: Adapt to the special needs of consumers in different parts of the world and produce specialized products needed.

Figure 6-12 Attention by various industries in internationalization of technological innovation to the market conditions of the host country

While most companies consider that the three indicators of internationalization of technological innovation are not important, some companies still think that they are the main factors influencing the implementation of their strategy for internationalization of technological innovation. We have found through the survey and interviews that there are two cases among such enterprises: the first case is that the enterprise has already reached a high level of technological development and a high level of international market penetration. Such enterprises—through internationalization of technological innovation of the technology learning type and the resource acquisition type—have already possessed a level of technological prowess; they have already set their goals on the markets of third world countries and have established their production bases overseas, and internationalization of technological innovation of the market support type has already become part of their internationalization of technological innovation. Huawei's Russian institute, for example, is for occupying Russia's wireless communications market; it was set up with the backing of Beto-Huawei joint venture in Russia and was set up especially for research related to the localization of Huawei technologies. The second case is in export-oriented enterprises, such as Shandong Luneng Institute, whose markets are mainly in India and the Middle East; close ties with local customers is the key factor for their internationalization of technological innovation. The common characteristic of internationalization of technological innovation of such enterprises is that they are closely linked with the host country's markets; thus, such enterprises pay more attention to the market conditions of the host countries.

In contrast, because most Chinese enterprises operate for the domestic market and thus have limited need for internationalization of technological innovation of the market support type, a host country's market condition is then not an important factor in their effort on internationalization of technological innovation.

(3) Host country's cultural differences being a barrier to internationalization of technological innovation for Chinese enterprises

As listed in Table 6-2, spatial, cultural and language barriers are also important factors affecting the decision-making of Chinese enterprises regarding internationalization of technological innovation. In China, 40% of the enterprises believe that this indicator is the most important or important; such enterprises are mainly concentrated in the high-tech industry (Figure 6-13).

Table 6-2 Evaluation of the impact of host country's cultural environment and government policies on internationalization of technological innovation for Chinese enterprises

Environment/ Policy	Average score	Most important	Important	Medium	Not important	Least important
Spatial, cultural and language barriers	3.1	20%	20%	30%	10%	20%
Foreign trade policy	3.71	21.4%	42.9%	21.4%	14.3%	0
Taxation, financial, and fiscal policies	3.36	21.4%	21.4%	35.7%	14.3%	7.2%
S&T and innovation policy	3.36	35.7%	7.1%	28.6%	14.3%	14.3%
Intellectual property rights policy	4.29	35.7%	57.1%	7.2%	0	0

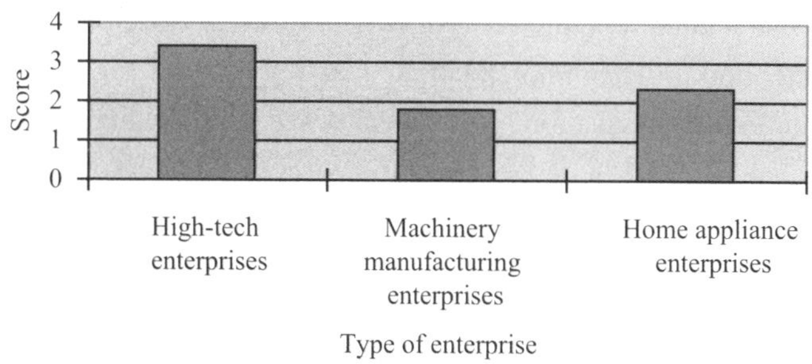

Figure 6-13 Impact of cultural barriers on decision-making in enterprise internationalization of technological innovation

High-tech industry is a knowledge-intensive industry; its R&D activities need close collaboration and communication, and its overseas R&D activities are mainly driven by the overseas R&D supply factors and thus belong to "active" R&D internationalization. Spatial, cultural and language barriers have large impact on the coordination of R&D activities and the communication among people; 30% of the enterprises do not think that they are important; such enterprises mainly carry out R&D internationalization of the market support type, and their overseas R&D activities are mainly pushed by the overseas market factors and thus belong to "passive" R&D internationalization. Having its own market abroad, supply-and-demand relationship with the local market, and accumulation in international marketing experience are the factors that will help an enterprise overcome spatial and cultural barriers.

(4) Host country's government policies have great impact on internationalization of technological innovation for Chinese enterprises

As listed in Table 6-2, Chinese enterprises gave higher average scores to the effect of host country's government policies than that of cultural environment, indicating that those government policies have a greater impact on internationalization of technological innovation for Chinese enterprises.

The host country's intellectual property rights policy, in particular, is the host country's policy factor to which Chinese enterprises will pay most attention to when implementing a strategy for internationalization of technological innovation. This indicator received the highest average score of 4.29; 92.8% of the enterprises think that this factor is the most important or important. On the one hand, this shows that Chinese enterprises have a heightened awareness of intellectual property rights; on the other hand, the type of R&D internationalization activities conducted by these enterprises play a significant role. As mentioned earlier, R&D internationalization activities by Chinese enterprises mainly involve technology search and technology learning; in the process of technological innovation, imitative innovation accounts for a larger proportion and, thus, the issue of intellectual property protection cannot be avoided. How to comply with intellectual property laws and how to act in accordance with international rules are both important questions for Chinese enterprises in the process of implementing R&D internationalization strategy, and strengthening the protection of the fruits of technological innovation will play a positive role in promoting the Chinese R&D internationalization process.

The host country's foreign trade, taxation, financial and fiscal policies, and S&T and innovation policies are also important environmental factors for Chinese enterprises in the process of internationalization of technological innovation. Because a substantial part of internationalization of technological innovation by Chinese enterprises involves international technology transfer, the part of foreign trade policies related to technology transfer constitutes a key factor for Chinese enterprises pursuing internationalization of technological innovation. At the same time, some countries consider technology transfer as an additional condition for market entry; such practice, for China's leading enterprises such as Hisense, Huawei, and ZTE, is on the focus of attention when carrying out internationalization of technological innovation of the market support type. The issue of costs is an important restrictive factor for Chinese enterprises when carrying out internationalization of technological innovation, and thus, the enterprises are sensitive to the host country's taxation, financial, and fiscal policies. A host country's S&T and innovation policies directly affect its technological innovation environment and the direction of its S&T development. For international R&D activities of the technology search type conducted by Chinese enterprises, a host country's S&T and innovation policies affect their information sources and the innovation processes—which thus should also be considered as a major factor.

6.4.2 Analysis of the impact of government policy adjustment on enterprises' internationalization of technological innovation upon China's accession to the WTO

After joining the WTO, in order to comply with the requirements stipulated by the WTO agreement, China could make some policy adjustments in accordance with WTO regulations. Through our survey, we found that these policy adjustments could have various levels of impact on enterprises' internationalization of technological innovation.

(1) Greater impact of strict enforcement of provisions related to the protection of intellectual property rights on enterprises

The biggest impact on enterprise technological innovation activities could be that most of the imitative innovations could probably be stopped and, as a result, enterprises should increase their R&D investment in order to obtain technological achievements—for which they own the intellectual property rights—through independent innovation. For Chinese enterprises that usually have tight funding, this will put tremendous pressure on the funding of R&D activities.

After joining the WTO, the enforcement of provisions related to intellectual property rights in China could further increase the monopoly of developed countries in knowledge supply; however, 35.7% of the enterprises believe that China would not necessarily have to pay more for acquiring foreign technologies and that developed countries would not necessarily slow down their technology transfer to China (Figure 6-14).

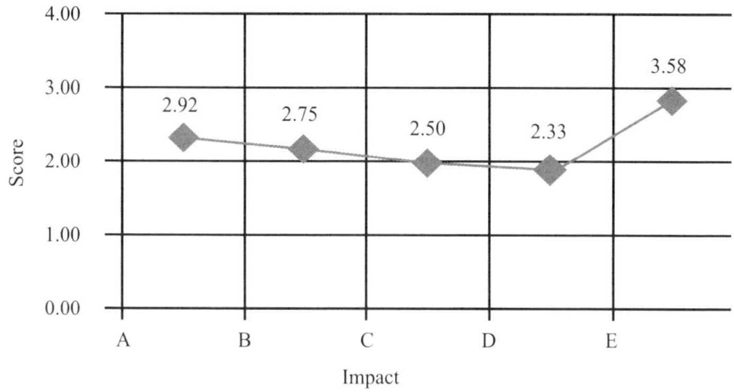

A: Protection of intellectual property rights could further increase the monopoly of developed countries in supply of knowledge;

B: Chinese enterprises would have to obtain foreign technology at a higher cost;

C: Developed countries slow down their technology transfer to China;

D: To an extent, prevented repetitive development of software products that are functionally the same and prevented the implementation of reverse engineering;

E: Prevented much of the imitative innovation, and as a result, China has to strengthen R&D to obtain its own intellectual property rights and Chinese enterprises are under tremendous funding pressure of R&D activities.

Figure 6-14 Impact of China's accession to the WTO on technological innovation by Chinese enterprises

(2) Negative impact of enforcement of the agreement on subsidies and countervailing measures on internationalization of technological innovation carried out by Chinese enterprises

In accordance with the terms of the WTO, China could enforce the agreement on subsidies and countervailing measures. The agreement prohibits various subsidies conditioned upon export performance and import substitution and, at the same time, strictly limits various subsidies for enterprise R&D, with an exception of basic research, industrial research, and precompetitive development activities for subsidies. The agreement also stipulates that subsidies for industrial research and precompetitive development should not exceed 75% and 50% of the reasonable cost, respectively.

Thus, lots of China's S&T and financial policies should be adjusted accordingly, and policies that are not in compliance with the WTO agreement, such as government grants, loan subsidies, VAT, export tax rebates, subsidies, and financial and fiscal incentives could be canceled. For example, preferential treatments to be canceled include the following: technology export tariff reliefs and export rebates; fiscal funding, subsidized loans, tax benefits, etc. for various R&D activities for import substitution; fiscal funding, subsidized loans, tax benefits, etc. for pilot tests, industrial exemplary production, and industrial experimental base; fiscal funding, subsidized loans, tax and depreciation benefits, etc. for enterprises' new product trial production, intermediate test costs, commercial production, development at market introduction stage, etc.; grants and fiscal funding for marketing phase development of national key new products and new product industrialization; and fiscal grants, subsidized loans, tax benefits, etc. for localization of equipment development that is important to the country.

At the same time, China could enforce the TRIMs agreement (the agreement on Trade-Related Investment Measures) and cancel the local content requirements, balanced trade requirements, domestic sales requirements, technology transfer requirements, foreign exchange restrictions, etc., which are instituted by China for foreign businesses.

Our survey indicated that all such policy adjustments could more or less have a degree of negative impact on Chinese enterprises.

6.4.3 Policy suggestions

The Chinese government should understand the characteristics of internationalization of technological innovation of Chinese enterprises and the environmental factors that they face, take effective measures, actively support them in their efforts to implement their strategies for internationalization of technological innovation, and create good environmental support for such efforts, so as to raise the overall technological level of Chinese enterprises, strengthen their international competitiveness, and improve the overall technological development of China.

Presently, China has not yet established a set of sound policy rules and a management system for enterprises' internationalization of technological innovation, such as special preferential support policies, information advisory and service institutions, and so on. For this reason, we believe that the government should take the following measures.

(1) When formulating the national S&T development policies and constructing the national innovation system, the government should clearly set out the principles and methods for guiding, supporting, and encouraging enterprises' internationalization of technological innovation

(2) Within the framework of WTO agreement, in addition to supporting enterprise activities for internationalization of technological innovation through financial and fiscal policies, a venture capital mechanism for such activities should be set up

Our survey showed that 53.9% of the enterprises believe that high cost is an important factor inhibiting their internationalization of technological innovation. The actual situation is that investment in R&D by Chinese enterprises is generally lacking when compared with the investment by foreign enterprises, and the cost for setting up an R&D unit abroad is more expensive than setting up an R&D unit in the home country. Thus, funding problem is an important factor inhibiting Chinese enterprises from implementing their R&D internationalization strategies. Venture capital intervention and capital market financing can effectively solve the problem of resource input shortage of activities in internationalization of technological innovation by Chinese enterprises.

(3) The government should increase subsidies for the purchase of patents and other soft technologies and provide a favorable environment for implementing the activities of internationalization of technological innovation that are mostly imitative innovation. At the same time, it should increase the support for enterprise patent applications and protection of trademarks and lead enterprise innovation activities gradually towards independent innovation

(4) The government should provide good public services and create favorable conditions for internationalization of technological innovation

The government should create and foster a favorable environment for internationalization of technological innovation and provide good public services, with the following steps.

1) Establishing information service centers and providing overseas technology and market information

Because the internationalization level of business operations of Chinese enterprises is not high, most enterprises lack a good understanding of the international market, the international environment, the social and cultural environment, and government policies of different countries. Hence, the information advisory services provided by the government can greatly promote enterprise internationalization of technological innovation.

2) Providing personnel education and training for enterprises' internationalization of technological innovation

The key factor to influence technological innovation is identifying the talent; high-quality technology and managerial talent in internationalization of technological innovation must be able to bridge cultural differences and have good communication skills and adaptability. Presently, most Chinese enterprises lack talent in internationalization, and the costs for overseas talent are very high. The government's help in training talent for internationalization will greatly promote enterprises' internationalization of technological innovation.

3) Establishing an intermediary service system for international cooperation in technological innovation and providing a bridge for technology cooperation between Chinese and foreign enterprises

Finally, through international agreement, Chinese enterprises should seek preferential policies in the host country for enterprises' internationalization of innovation activities, so as to provide a good environment for enterprise technological innovation activities in the host country—for example, the signing of bilateral treaties with the host country for protecting business investment to avoid confiscation or expropriation of investment assets and the signing of international tax agreements to avoid double taxation between the two countries.

6.5 Chapter summary

This chapter, from the environmental support angle of internationalization of technological innovation, analyzed the mechanism by which the environmental factors of the host country and the home country affect an enterprise and its internationalization of technological innovation and also analyzed the main environmental factors to be considered by Chinese enterprises in the current stage of internationalization of technological innovation. We believe that good environmental support is the external guarantee for the smooth development of enterprise internationalization of technological innovation. In response to the reality being faced by Chinese enterprises after China's accession to the WTO, the Chinese government should take effective policy measures to actively support Chinese enterprises in the implementation of their strategies for internationalization of technological innovation and to create good environmental support for such internationalization strategies, so as to raise the overall technological level of Chinese enterprises, strengthen their international competitiveness, and improve the overall level of China's technological development.

References

Bartlett, C. A. & Ghoshal, S. *Managing Across Borders: The Transnational Solution*. Boston: Harvard Business Review Press, 2002.

Cantwell, J. A. *Technological Innovation and Multinational Corporations*. Oxford: Basil Blackwell, 1989.

Cantwell, J. A. The globalization of technology: what remains of the product cycle model? *Cambridge Journal of Economics*, 1995, 19(1): 155-174.

Cantwell, J. A. & Piscitello, L. Accumulating technological competence: its changing impact on corporate diversification and internationalization. *Industrial and Corporate Change*, 2000, 9(1): 21-51.

Cantwell, J. A. & Kosmopoulou, E. Determinants of internationalization of corporate technology (DRUID working papers No 01-08). Copenhagen: DRUID, 2001.

China Society of Science and Technology Indicators. OECD Analytical Method and Empirical Study of R&D Data, 2002.

Dunning, J. H. *Multinationals, Technology and Competitiveness*. London: Unwin Hyman, 1988.

Granstrand, O., Hakanson, L. & Sjolander, S. *Technology Management and International Business: Internationalization of R&D and Technology*. Chichester: Wiley, 1992.

Kumar, N. Determinants of location of overseas R&D activity of multinational enterprises: the case of US and Japanese corporations. *Research Policy*, 2001, 30(1): 159-174.

Lundvall, B. -Å. *Product Innovation and User-Producer Interaction*. Aalborg: Aalborg University Press, 1985.

Lundvall, B. -Å. Innovation as an interactive process: from user-producer interaction to the national innovation systems. In Dosi, G. et al. (eds.), *Technology and Economic Theory*. London: Pinter Publishers, 1988.

Moenaert, R. K., Caeldries, F., Lievens, A. & Wauters, E. Communication flows in international product innovation teams. *Journal of Product Innovation Management*, 2000, 17(5): 360-377

OECD. Measuring globalisation: the role of multinationals in OECD economies. Paris: OECD, 1999.

Chapter 7

Conclusion

In the course of this project, we studied the developmental patterns of internationalization of technological innovation by Chinese enterprises in three aspects—strategy, organizational form, and resource allocation. We also analyzed the support environment for internationalization of technological innovation. Based on the systematic analysis of the actual situation of internationalization of technological innovation by Chinese enterprises, we summarized the evolutionary patterns and management features of such internationalization and put forward some policy suggestions.

7.1 Strategic aspect of internationalization of technological innovation

We have proposed a strategic framework for internationalization of technological innovation that is based on core competency. The strategic objective of this framework requires that Chinese enterprises achieve globalization in the medium- to long-term future in both the technology and the market dimensions. In other words, Chinese enterprises have to extend their technological innovation activities to every corner of the world, diversify their sources of technological knowledge, truly obtain the knowledge and technology required for technological innovation from the world's various knowledge and technology centers of excellence, and enhance their core competencies through internationalized technological innovation activities, so as to move toward internationalized production and operations and ultimately achieve globalization of their production and operations as well as technological innovation activities.

We believe that in this direction, during different developmental stages, the strategic path for internationalization of technological innovation by Chinese enterprises will go through different and sequentially evolutionary forms as follows.

(1) Preliminary stage of internationalization of technological innovation

The enterprise focuses on the domestic market, identifies relevant innovative resources through technology monitoring, and concentrates its R&D activities in the home country. This will be adopted as the main strategic path at this stage. This stage is characterized by the domestic market, domestic R&D activities, and overseas technology monitoring points, with the purpose of acquiring information about technology development worldwide and using it as the source of idea for technological innovation.

(2) Initial stage of internationalization of technological innovation

The enterprise focuses on the domestic market, but with an eye on the international market; its R&D activities are concentrated in the home country, but at the same time the enterprise starts to set up overseas R&D units of the technology monitoring type and carry out extensive international cooperation, so as to enhance the enterprise technological capabilities. At this stage, this is often adopted as a strategic path for technological innovation. The goal at this stage is to move toward internationalization in both the technology and the market dimensions. However, the strategic emphasis is still on domestic R&D and domestic market. By setting up overseas R&D units of the technology monitoring type and by engaging in extensive international cooperation, at this stage the enterprise achieves the strategic objectives of understanding competitors, determining technological trends, absorbing advanced technologies and scientific knowledge domestically and from abroad, enhancing technological capabilities, and, on such a foundation, strengthening its abilities to exploit the international market. Location selection for an overseas R&D unit is mainly influenced by the technological innovation environment and the availability of S&T resources in the host country, which usually results in choosing a place near a technology center of excellence.

(3) Development stage of internationalization of technological innovation

Setting the focus on both the domestic and international markets, establishing R&D branches from scratch or acquiring R&D branches via M&As, and thereby moving toward the exploitation of the international market with overseas R&D activities constitute the strategic approach and direction of this stage.

The strategic emphases are, on the one hand, to get close to the world's advanced innovative environment, take advantage of the spillover effect of technological innovation, absorb advanced technical knowledge, and extend and enhance potential technological capabilities of the enterprise and, on the other hand, to carry out technological innovation based on the conditions of the international market and provide support for the enterprise's overseas production and operations. The domestic R&D center still takes a very important central position. The enterprise directs R&D activities at overseas R&D branches through its domestic R&D centers. In the process of selecting a location for an overseas R&D unit, market factor of the host country is also an important parameter.

(4) Mature stage of internationalization of technological innovation

Internationalized and networked R&D with a focus on the international market is the strategic approach and the ultimate development goal of this stage.

With this approach, technological innovation is directed at the international market and international technology, and there is no central position for the domestic R&D center as in the past. All the R&D institutions are integrated into a network that becomes a relatively unified and coordinated network organization. In such a networked organization, all knowledge and technology can be shared. Each R&D branch operates at a higher level of independence, and its initiative and enthusiasm can be brought into relatively full play. Such an organizational structure is conducive to greater efficiency in R&D activities. To integrate the various overseas R&D units is the central task of location selection—in accordance with the enterprise's overall business strategy, technology strategy, and environmental factors of various host countries.

7.2 Organizational aspect of internationalization of technological innovation

We believe that with strategic development, the organizational structure for internationalization of technological innovation of Chinese enterprises should be constantly adjusted.

(1) Initial stage of internationalization of technological innovation

The organizational system for internationalization of technological innovation of Chinese enterprises is generally of the center-edge structure.

Such structure has the R&D center of the company's headquarters at the center and a number of overseas R&D institutions at the edge. In this organizational structure, the R&D center of the headquarters carries out the innovative activities and the main functions of overseas R&D units are technology search or secondary development. Decisions are mainly made by the headquarters, overseas R&D units are responsible for the implementation, and exchange and communication within the organizational structure are done through formal channels of a hierarchy. Overseas R&D units are coordinated by the R&D center of the headquarters, and the contact among the overseas R&D units is minimal.

(2) Development stage of internationalization of technological innovation

Overseas R&D units not only engage in technology search but also work on technology transfer or on product development for the market serviced by a subsidiary there. As the function of overseas R&D units changes and the overall organizational system becomes more complex, the pure hierarchical star structure and the formal communication channels can no longer adapt to strategic development. It becomes necessary to establish a committee directing international innovation activities and operating concurrently with the hierarchical star structure. In such a structure, the committee operating in parallel not only plays an important role in innovation activities but also provides a platform for information transfer within the enterprise. The branch unit sheds the role as a mere executor and gains a great deal of autonomy. Formal and informal communication channels coexist, and informal channels promote communication among the branches.

(3) Mature stage of internationalization of technological innovation

In this stage, the organizational system turns into a networked structure. Various R&D units are set up differently in accordance with the different strategic positions of their environment and their different specializations. Within the overall R&D system, different units are dependent on each other. There are no central innovation centers, and the headquarters and various R&D units work together to formulate a complex process of coordination and cooperation within the environment. Among the mutually dependent units, there are substantial flows of people, information, and other resources.

7.3 Resource allocation aspect of internationalization of technological innovation

Adapting to the development in the matching mode of the strategy—organization pair, the resource allocation mode of the internationalized technological innovation system will also be adjusted so as to maintain the agreement among the business strategy, organizational form, and resource allocation scheme.

(1) Initial stage of internationalization of technological innovation

Major resources are concentrated in the company's headquarters; human and financial resources are channeled mainly from the headquarters to overseas R&D branches; and overseas R&D branches send the information and knowledge they acquired about advanced technology to the headquarters of the home country.

(2) Development stage of internationalization of technological innovation

Most resources are still concentrated in the headquarters, but they have become noticeably dispersed, and overseas R&D branches, depending on their different strategic positions, are in possession of unequal resources. The means for resource acquisition by the overseas R&D branches are no longer limited to the inputs mainly from the parent company, the local recruitment and interunit personnel flow have greatly increased, and the sources of capital input have diversified—headquarters, overseas subsidiaries, venture capital, and various noncommercial organizations can be investing parties. Information exchange has changed the flow from unidirectional between the layers of organizational hierarchy to bidirectional, thus leading to the existence of information exchange among the branch units.

(3) Mature stage of internationalization of technological innovation

Resources are no longer concentrated in the headquarters, and they are allocated among the R&D units within the networked system in accordance with their varied levels of strategic importance. The approaches to acquiring resources by the overseas R&D units have been diversified. The local recruitment of human resources and their interagency flow are the main characteristics; information sharing has been achieved, and the pattern of information flow is now multidirectional.

7.4 Managerial aspect of internationalization of technological innovation

Internationalization of technological innovation has posed substantial challenges to the traditional technological innovation management—because of the globalized acquisition of innovative sources, the networked form of innovative organization, and the multichannel nature of innovative resource configuration.

To formulate a strategic vision of an open system for internationalization of technological innovation is the most important task for managing internationalization of technological innovation. The essence of conventional technological innovation—regardless of where the information is from, namely, internal R&D units, customers, and suppliers—is still innovation that is carried out internally within an enterprise. In order to take the initiative, make use of worldwide technological innovation resources, and carry out innovation externally, the enterprise must change the mode of closed technological innovation and formulate an open strategic vision of international technological innovation.

Successful localization of overseas R&D units is the basis upon which the overseas R&D units develop. Because the environmental factors faced by an overseas R&D unit, such as social and cultural factors, political and economic factors, and technological innovation resource factors, are all quite different from those of the home country, for the unit to overcome communication difficulties and organically integrate into the local environment is the key factor for successful technological innovation. Thus, recruiting local talent, strengthening cultural communication, and achieving localization of overseas R&D units constitute the foundation of the development of overseas R&D units.

Because of their dispersed geographic locations, different positions in the organizational system, and different environmental conditions, the various R&D units may have conflicting interests, incongruent goals, different operating modes, and difficulties in communicating with each other. Thus, taking the view of the enterprise as a whole, functionally consolidating those geographically dispersed units, strengthening communication and cooperation among the units, reconstructing the innovative process and maintaining the congruity and agreement between business strategy, organizational form, and resource allocation scheme are the focal points of efforts to preserve and improve the efficiency of technological innovation.

7.5 Evolutionary modes and management features of China's internationalization of technological innovation

After analyzing the strategic, organizational, and resource configuration aspects, we put forward the evolutionary modes and management features of internationalization of technological innovation for Chinese enterprises as listed in Table 7-1.

Table 7-1 Evolutionary modes and management features of internationalization of technological innovation for Chinese enterprises

		Initial stage	Development stage	Mature stage
Strategy	Motivation	Determining technology trends, absorbing advanced technology and scientific knowledge from abroad, and enhancing enterprise technological capabilities	On the one hand, get close to the world's advanced innovative environment, absorb advanced technology and scientific knowledge, and extend and enhance potential technological capabilities of the enterprise; on the other hand, carry out technological innovation based on the conditions of the international market and provide support for the enterprise's overseas production and operations	Make use of global innovative resources and environment, draw near to the large markets, engage in global innovations, foster the interplay of technological innovation and market exploration, and fully enhance the potential technological capabilities of business enterprises as well as increase their share of the world market
	Positioning	Mainly influenced by the technological innovation environment and the availability of S&T resources in the host country, usually resulting in being located at a place near a technology center of excellence in the world	Influenced by the technological innovation environment and the availability of S&T resources, with host country's market factor also being a major parameter	In accordance with the different location advantages as a result of the host countries' different technological innovation environment, S&T resources, and market condition, reasonably allocate R&D resources
	Form	R&D activities are concentrated in the home country, but at the same time enterprises start to set up overseas R&D units of the technology monitoring type and carry out extensive international cooperation	R&D activities are mainly concentrated in the home country, and overseas R&D activities of the resource utilization type, technology acquisition type, and market support type coexist	There are no unique R&D centers; R&D activities are rationally arranged around the globe based on the enterprise strategic emphasis and the host country's location advantages

(Continued)

		Initial stage	Development stage	Mature stage
Organizational form		Center-edge structure—with the R&D center of the headquarters at the center and a number of overseas R&D institutions at the edge. The R&D center of the headquarters carries out innovative activities, and the main functions of overseas R&D units are technology search or secondary development	Composite-star structure—headquarters is the center for strategic decision-making, and R&D activities are decentralized. Set up an international steering committee that operates concurrently with the star-shaped hierarchical structure, and carry out project coordination across hierarchical levels	Globalized networked structure—there are no central innovation centers, various R&D units are set up differently in accordance with the different strategic positions of their environment and their different specializations, and within the overall R&D system, different units are dependent on each other
Resource allocation	Information	Unidirectional flow from the branches to the headquarters	Mainly bidirectional flows between the branches and the headquarters, with intensifying information exchanges among various units	Information sharing and multidirectional information flows
	Human resources	Mainly dispatched from the headquarters	Local recruitment and cross-unit mobility	Global recruitment and substantial cross-unit mobility within the enterprise
	Funds	Input by the headquarters	Diversification of input types	Diversification of input types
Management		Decisions about setting up overseas R&D goals, planning overseas R&D projects, and so on made at the headquarters; branches responsible for the implementation; and formal channels for exchanges	Strategic decisions made at the headquarters; branches having more autonomy; and formal and informal channels of communication coexist	Functionally consolidating those geographically dispersed R&D units; strengthening the communication and cooperation among the various units; reconstructing innovative processes; and maintaining the congruity and agreement between business strategy, organizational form, and resource allocation scheme are the focal points of effort to preserve and improve the efficiency of technological innovation

7.6 Environmental support conditions of internationalization of technological innovation for Chinese enterprises and policy implications

Our research has led us to the belief that the host country's technological innovation resource condition is presently the main factor in the focus of attention for Chinese enterprises when carrying out internationalization of technological innovation; the host country's market condition is not the main environmental factor of such internationalization; the difference between the cultures of the host country and the home country is a hindering factor; and the host country's government policies have a relatively large impact on such internationalization.

After joining the WTO, in order to comply with the requirements stipulated by the WTO agreement, China has made some policy adjustments in accordance with WTO regulations; these policy adjustments have had various levels of negative impact on the original mode of internationalization of technological innovation for Chinese enterprises.

Presently, China has not yet established a set of sound policy rules and a management system for enterprises' internationalization of technological innovation. For the Chinese government, taking effective measures to actively support enterprise technological innovation activities and creating good environmental support for the effective implementation of enterprises' internationalization of technological innovation have great significance for raising the overall technological level of Chinese enterprises, which help the enterprises to strengthen their international competitiveness and improve the overall level of China's technological development.

Appendix 1

Case 1 Motorola's innovation internationalization

Motorola is the industry leader in global electronic communications. The company was created in 1928 by Paul V. Galvin in Chicago; at that time, it took the name "Galvin Manufacturing Corporation." In the 1930s, the company successfully started to operate in the car radio business with the Motorola brand and then started to produce home radio and police radio equipment, with a lot of advertisements. In 1947, the company changed its name to Motorola. When Galvin died in 1959, Motorola had already become a leader in military, aeronautical, and commercial communications. Under the leadership of Galvin's son, Robert W. Galvin, the company started international marketing in 1960, with a focus on consumer electronic products. Toward the end of the 1980s, Motorola became the world's largest cellular phone provider, and in 1996, it began producing portable wireless phones. In addition, merging General Instrument Corporation with Motorola also made Motorola the leader in cable modem and set-top box.

With the rise of knowledge economy and international competition, the original centralized parent company's R&D is no longer suitable. On the one hand, knowledge economy-related resources exist across national boundaries, and in order to obtain new knowledge for multinationals as well as to acquire advanced research products from overseas universities and competitors, they have to set up R&D units in areas close to universities and to their competitors. On the other hand, competition in the international market forces multinationals to speed up the process from the new product development stage to the market introduction stage. As a result, multinationals, in order to stay competitive, must set up international R&D networks. Such networks can not only help them acquire information from new knowledge centers but also help them commercialize products in the overseas markets. In view of this, Motorola has set up research institutes in several countries; such institutions carry out R&D projects in cooperation with overseas multinationals or research institutes. In the following sections, we use China as an example to illustrate Motorola's R&D internationalization process.

1. Profile of Motorola China

Motorola China was established on November 3, 1999; it is under the leadership of a board of directors. In 2001, Motorola's R&D investment in China reached $300 million. Ever since its establishment, its China Institute has achieved success in promoting technology exchanges and technology sharing among various research centers and even across the entire Motorola (China) Electronics Ltd. For the career development of various R&D personnel, the institute has actively promoted the Technology Ladder project. In order to protect the inventions and innovations of Motorola, the institute has actively promoted patent application within the company and has set up a patent committee; meanwhile, the institute has gradually strengthened friendly exchanges with various social organizations and has strived to open up new cooperative projects and partnerships in addition to maintaining collaborative relationships with existing partners.

On November 27, 2001, Motorola's global board of directors convened in Beijing and decided to build one of Motorola's global R&D bases in Beijing—which will house core technology research—to invest a total of $1.3 billion in R&D in the following five years, and to have more than 5,000 R&D personnel.

Motorola's China Institute will continue to pursue long-term success in China and to maintain its leading position in the fields of telecommunications and semiconductor through technology development and innovation. The institute will combine the technological strength and advantages of the R&D centers, so that Motorola (China) Electronics Ltd. will have the ability to provide integrated technology solutions from chip development to system design and to application development around the globe.

Motorola's four business principles for its China business are as follows:

(1) Investment and technology transfer;
(2) Localization of talent management;
(3) Localization of accessory products;
(4) Joint venture and cooperation.

2. Technology transfer and localization—main R&D center

(1) Beijing R&D Center of Personal Telecommunications Division

In March 1999, in order to enhance its R&D strength in China, Motorola Communications Group merged the original Motorola China Mobile Division R&D Center and the Pager Division R&D Center and established Beijing R&D Center of Personal Telecommunications Division. This center is one of the R&D centers in the Asia-Pacific region set up by Motorola Personal Communications Division. Every year, Motorola invests on average tens of millions of USD in project development, equipment procurement, and personnel training. The mid- to high-level R&D personnel in the research center are all engaged in R&D in software, electrical, mechanical, industrial design, and other fields that are related to personal communications products. The center has completed a number of product R&D projects, including technology for Chinese character input on mobile phone keyboard, Motorola's innovative WAP mobile phone A6188, and so on.

(2) Applied R&D Center for the Asia-Pacific Region of Telecom Solutions Division

Motorola's China Network Solutions Technology Center (hereinafter referred to as the Technology Center) of the Global Telecom Solutions Division is one of Motorola's global development centers.

The Technology Center was established in October 1995 in response to the need of China's telecommunications market. It has developed into a center that provides communications network solutions to the Asia-Pacific region and even to the entire globe. The headquarters of the Technology Center is set in Beijing, and a branch unit was set up in Chengdu in 1999. In 2000, the Technology Center received the SEI CMM (Capability Maturity Model) Level 5 certificate from the Software Engineering Institute of Carnegie Mellon University, which confirms that the Technology Center has already attained the ability to complete high-efficiency and high-quality product development; this achievement has been highly appreciated by Motorola.

(3) Software R&D Center of Professional Wireless Communications Division

Established in March 1998, Software R&D Center of Professional Wireless Communications Division is committed to developing cutting-edge wireless walkie-talkie products, and its engineers work with the best engineers at Motorola design centers in the United States, Europe, and Malaysia to provide solutions for Motorola's Asian customers. In line with the strategy of common software platform in product development, the center will continue to meet new challenges, develop new products, and help Motorola's CGISS (Commercial, Government, and Industrial Solutions Sector) customers to improve the operating performance of their products.

(4) Motorola China Software Center

Motorola China Software Center (hereinafter referred to as the Software Center) was established in 1993; it is an important part of the Motorola Global Software Group. The center is the first software development institution in the Chinese mainland, which carries out its business on the basis of the CMM for Software. This model divides software development institutions into five levels, with Level One being the lowest and Level Five being the highest. The Software Center passed CMM Level Five accreditation in September 2000. Since 1993, the Software Center has been working on providing customers with the best software solutions and has achieved total customer satisfaction.

(5) China Semiconductor Technology Center of Motorola Semiconductor's Digital Genetics Laboratory

China Semiconductor Technology Center of Motorola Semiconductor's Digital Genetics Laboratory was established in 1995; it is an important laboratory in Motorola's global semiconductor technology research. The main areas of its research are in semiconductor devices and process simulation and prediction technologies, the development and integration of new technological processes, system chip integration technology, advanced semiconductor materials and related software technology research, and so on. On December 1, 2000, the center received its CMM Level Three certificate and thereby became the first semiconductor unit in China to have been certified as a CMM3 institution.

(6) Semiconductor Design Centers

In order to grasp the pulse of China's S&T development, to adapt to the rapid expansion of the market for high-tech products, to contribute to the company's strategy for long-term development in China, and to get closer to customers and provide them with better services, Motorola (China) Electronics Ltd. brought its high-level microcontroller product design technology to China and set up the Motorola Suzhou Semiconductor Design Center in Suzhou New District. The first implemented projects include wireless communications and microcontroller chip design. On August 23, 1999, Motorola Suzhou IC Design Center was officially established. The center is known as China's largest national integrated circuit design center. It has the most advanced hardware and software environment and is engaged in the R&D and design of high-end products.

Another institution of Motorola's semiconductor design is its Hong Kong Design Center. Hong Kong Design Center works on the 8-/16-/32-bit microcontrollers and other CMOS technological processes and designs products for consumer applications and specific automotive applications.

(7) Embedded Systems Development Laboratory

The Embedded Systems Development Laboratory (ESDL) was established in January 2000. Its establishment marked the successful completion of the division's smooth transition from Motorola-ICT Advanced Human-Machine Communication Technology Joint Laboratory in Beijing to ESDL. ESDL's missions are to make use of its advantage in technology support and to provide ancillary software modules for various PowerPC-based platforms, so as to strengthen the position of Motorola's semiconductor business in China and promote the growth of the related market in China.

ESDL's R&D is based on the Internet and the PowerPC field. In the area of technical support, it provides answers to customer inquiries and solutions to electronic mail hotlines in China. In the field of software development, it is committed to developing cutting-edge technologies—especially those catering to the need of the Chinese market.

(8) Motorola's Advanced Technology Center

Motorola's Advanced Technology Center (MATC-Asia) was established in 1995, which is located in Tianjin Economic-Technological Development Area. This center has a modern materials research laboratory and a software development laboratory; it has materials and failure analysis and CIM software R&D capabilities. Its goal is to become a first-class technology center in materials analysis within the field of electronics manufacturing and assembly in the Asian region.

The basic functions of the center are to support Motorola's production operations in the Asia-Pacific region, providing technical training and assisting in the import and implementation of advanced technologies; to conduct materials analysis services, providing solutions; to assist in the development and implementation of CIM technology in Motorola; and to cooperate with the local research institutes in R&D projects.

(9) Motorola China Research Center

Motorola China Research Center is located in Shanghai and was established in 2000. It is the branch unit in China of the Motorola laboratory under the company's headquarters and is its fifth overseas research center in the world. Motorola China Research Center is committed to medium- to long-term research on technologies related to the regional standards of China and to Chinese language and culture, so as to support the mainstream products that Motorola will provide in the Chinese market in future years. Its research projects are concentrated in various human-computer interaction technologies, such as embedded Chinese speech recognition and speech generation, Chinese natural language processing, Chinese character recognition, and multimodal technology for China's third-generation mobile systems. It has set up four research teams that are engaged in research in areas such as voice recognition, speech synthesis, natural language processing, pattern recognition, and Chinese phonetics research.

Through these research centers, Motorola can recruit local talent in the field of communications. On the one hand, it can accomplish investment and technology transfer, so that each research center has international-level R&D capabilities and technologies; on the other hand, through talent localization, it can develop high-tech products needed by the market more rapidly, better provide services needed by the local customers, further expand Motorola's market share, and at the same time reduce management costs and increase management efficiency.

3. Technology transfer and localization—localization of ancillary products

There are several hundred local suppliers of ancillary products in China. The average localization ratio is 69%, and supplier training spans areas of quality, management, and technological processes, etc. The types of components provided are as follows: semiconductor devices, framework for integrated circuits, micro motors, liquid crystal display, multilayer circuit boards, and so on.

The localization of ancillary products and acquisition of high-quality components inside China are not only a promise that Motorola has made to China, but they are also part of Motorola's development strategy. Even as early as in 1992, at the time when Motorola first set up its manufacturing plant, the company had decided to establish close partnerships with China's most promising component suppliers and to support their development.

The unique resource localization strategy carried out by Motorola has achieved tremendous success in the following areas:

- Greatly helped the improvement of the technology and product quality of domestic suppliers;
- Increased their knowledge in technology and management;
- Injected new vitality into local suppliers;
- Promoted the establishment of joint ventures by the overseas suppliers and the best local suppliers for their common development;
- Helped China to develop its electronics industry;
- Saved and earned foreign exchange.

Motorola has not only purchased components locally to support its development in China, but it has also helped local suppliers to export components and materials to Motorola's overseas manufacturing units outside China.

Ancillary product localization strategy benefits both parties of the joint venture. There are multiple reasons for the huge investment that Motorola has made in ancillary product localization. The first is that it is a wise move economically. The prices for locally acquired materials are reasonable, which means lower production costs and higher customer benefits. Ancillary product localization also shortens the production cycle of Motorola products and improves production efficiency and quality. Purchases of components locally make Motorola work with the nearby cooperative partners and ensure easy communication and bidirectional feedback. More importantly, the localization of ancillary products benefits China.

For China, the localization of ancillary products is of great benefit to the economy and technological development. By promoting the localization of ancillary products, Motorola has injected new vitality into many Chinese national enterprises and has made them profitable. China's suppliers have praised Motorola for bringing stability and growth to their factories and thereby increasing the economic income of many workers. The Chinese government has always supported

Motorola in its effort to help local companies and communities and to inject economic vitality into them.

Motorola has established close cooperative relationships with the local suppliers, helping them improve their ability to manufacture high-tech products. Motorola establishes cooperative relationships with suppliers who are able to meet Motorola's high-quality standards in production, helping them formulate equipment upgrade plan, set up quality control system, and improve the level of management.

Motorola has also invested funds for carrying out technology and management training for its suppliers and laying out a targeted training program for each of its important suppliers. In the process, Motorola has utilized its training forces from Beijing, Tianjin, and other parts of the world.

Motorola has worked to promote the establishment of joint ventures by first-class foreign suppliers and the domestic suppliers with great potential. Through investment promotion, Motorola has helped the foreign suppliers to better understand the Chinese market and has recommended component suppliers in China to such foreign suppliers, so that they can form joint ventures and gain good technology or high efficiency through cooperation. State-owned and large- and medium-sized enterprises are the pillars of China's economy. Success of the state-owned enterprise reform is crucial for China in achieving its goals of modernization, achieving social stability and economic development, and implementing the grand policy of continued reform and opening-up to the outside world. As a direct beneficiary of China's reform and opening-up policy, Motorola is fully aware of the importance of China's state-owned enterprise reform. Motorola has regarded the revitalization of China's state-owned enterprises as its own responsibility and has made solid contributions to helping China's stated-owned enterprises to get out of difficulties.

On April 2, 1997, Motorola and the National Planning Commission signed a memorandum for jointly formulating a program that is aimed at supporting the effort to improve the productivity and quality management of China's state-owned enterprises. The program includes training, arrangement of teaching personnel, teaching methods, structure and composition, and so on. Of all the foreign enterprises in China, Motorola is the first in putting forward an explicit programmatic document for the development of China's state-owned enterprises. This is the beginning of the subsequently established "Motorola Optimization Center for State-owned Enterprises," which—as Motorola's institution for organizing, coordinating, and implementing the work related to providing training for China's state-owned enterprises—has done a lot of work and made remarkable achievements.

Motorola's partners fall into four business areas: network operations, manufacturing, R&D, and application. The partners in network operations include China Telecom, China Unicom (CDMA system), Netcom, and China Mobile (GSM mobile); those in manufacturing include Eastcom, Zhongmei, Huaxu, Leshan, SVA, and other communications equipment manufacturers; those in R&D include various universities, Little Swan, CASS, NFGK, Daxian, Datang, and Eastcom; and those in application include Jinpeng, Datang, Huawei, Alcatel-Lucent Shanghai Bell, NEU-ALPINE, and other large enterprises. R&D projects have high technology content, and thus, close contact with various universities is indispensable.

Case 2

Haier's internationalization of technological innovation: inexhaustible source of power

1. Technological innovation: Haier going from the brink of bankruptcy to the realm of glory

Haier was formerly known as Qingdao Refrigerator Factory. In 1984, when first founded, Haier Group was a collective enterprise that was insolvent and at the brink of bankruptcy; it had 600 employees and had an annual sales revenue of 3 million yuan and an annual loss of 1.47 million yuan.

(1) Technology introduction and absorption: Haier running at the beginning

The basic problem faced by the new director Zhang Ruimin was how to introduce new technology as soon as possible so as to improve product quality and lead the factory to find a way out. After comparing 32 foreign manufacturers, Haier decided to adopt the advanced technology of Liebherr of Germany and signed the contract for technology import. In October 1984, Qingdao Refrigerator Factory became the last designated refrigerator factory. From there, Haier Group started its journey of technological innovation.

Technological innovation of this phase began with understanding and absorbing the refrigerator production technology of Liebherr of Germany. For this purpose, Haier set up the Technology Division, which was tasked with digesting and absorbing new technology, and sent more than 40 core members in technology and management departments to Germany for training and on-site learning. By such measures, the company ensured that its production line installation and refrigerator production operate at the level of international standard and, at the same time, it gradually mastered the key technologies for refrigerator development, trained competent development talent, and accumulated valuable experience in technology management. In 1987, in view of the diversified market demand and its improved capabilities, the company decided to build up and transform the Technology Division into a financially independent Refrigerator Institute. This change has increased the speed of product development.

High-level technology imports and effective digesting and absorbing of such technologies have led the company to a rising path. In 1985, the company launched Asia's first-generation four-star refrigerator "Qingdao Liebherr"; in 1988, it was awarded the first gold medal in the history of China's refrigerator industry and became one of the country's 500 largest industrial companies.

(2) Independent innovation: Haier reaching the realm of glory

In order to increase its technology reserves and enhance its market competitiveness, Haier invested 160 million yuan in 1992 to set up Haier Group Technology Center. Haier Group Technology Center is built upon the base of Qingdao Refrigeration Technology Research Institute and has set up a ternary development system; in other words, a comprehensive research center at Level One, four basic technology applied research centers at Level Two, and development centers for specific products at Level Three. Among them, the comprehensive research center works on medium- to long-term core technology and basic technology for the group; the four basic technology applied research centers work on developing medium- to short-term new products and related key technologies, with the main focus on the application of technology; the development centers for specific products work on developing short-term new products and improving functions and lowering cost.

Ever since the technology center was established, the speed for new product development has increased rapidly. On average, a new product comes out every 6–7 days and a patent application is made every 4–5 days. Among the new products, frost-free replacement MSV refrigerator and BCD-212 refrigerator won the second prize of the National Science and Technology Progress Award, and super energy-saving fluoride-free refrigerator was recognized as internationally advanced (see Table 1).

Table 1 Research Units within the Technology Center

Name	Field of Research
Qingdao Refrigeration Technology Research Institute	CAD design, large laser system
Qingdao Haigao Design & Manufacturing Co., Ltd.	Product design, product planning, design and information
Production Process Development Center	R&D in home appliance-related technology and equipment, mold, plastic, sheet metal products
Plastic Mold Research Institute	Plastic mold R&D

In 1997, Haier Group was designated by the State Economic and Trade Commission as one of the first six pilot enterprises for technological innovation. Haier Group Technology Center has also become one of the first state-level technology centers, and its main goal is to become one of the world's top 500. In 2000, in the evaluation of China's 294 state-level enterprise technology centers, it was ranked No. 1; meanwhile, it was named by the government of Shandong Province as the example unit of "Haier-style Enterprise Technology Center."

In order to increase technology reserves and enhance their market competitiveness, Haier Group has invested 500 million yuan to build a Central Research Institute with first-class facilities, which mainly develops new products five to ten years ahead. Haier Group also stipulates that the company's annual funding for R&D must not be less than 4% of its sales revenue and that those not asking for enough or not using enough will be punished.

(3) Integrating S&T personnel and leveraging strength to ensure vitality in technological innovation

In order to provide continued driving force for technological innovation, Haier combined manufacturing, learning, and research; adopted approaches, such as joint development; and integrated various scientific resources for its use. In January 1998, for the first time in China's S&T industrialization process, Haier Group launched an all-new approach of research-business cooperation via capital investment—operating as a majority stock holder of a state-level research institute: National Engineering Research Center of Engineering Plastics. It has adopted Haier's management approach and culture to manage the center, which has greatly enhanced Haier's research capabilities in the field of basic and forward-looking technology related to engineering plastics.

In April 1998, for the second time Haier Group decided to operate as a majority stock holder of a state-level research institute—Haier Guangke Digital Technology Co., Ltd., a joint venture established with the Academy of Broadcasting Science of SARFT. In June 1998, Haier Group, together with Beihang University and C-MOLD company of the United States, established Northern Haier Software Co., Ltd. as a joint venture, by which Haier became the first company in China's home appliance industry to enter the software industry (see Table 2).

Table 2 High-tech R&D institutions and enterprises that Haier invested in or acted as a cooperative party

Name	Partner
National Engineering Research Center of Engineering Plastics (Haier Holdings)	National Engineering Research Center of Engineering Plastics
Haier Guangke Digital Technology Co., Ltd.	Academy of Broadcasting Science of SARFT
Haier Beihang Software Co., Ltd.	Beihang University and C-MOLD company of the United States

Haier has worked with 120 professors from 25 universities to have an established joint development network and has set up five postdoctoral centers jointly with Fudan University, Shanghai Jiaotong University, and Zhejiang University. It has carried out doctoral research in various subjects such as chemistry and materials science, digital technology, software technology, bioengineering, overseas local design, and so on. On the one hand, it can acquire high-tech talent urgently needed for business development, form teams on research subjects around such academic leaders, and carry out focused research in mainly forward-looking technologies; on the other hand, it can take advantage of the strong technology research capabilities of the universities, create market advantages from technology advantages, and at the same time train talent needed for the company.

(4) Good organizational system

Haier's technological innovation system has three aspects: the organizational system for technological innovation, the process management system for product development, and the incentive mechanism for technological innovation. The organizational system for technological innovation consists of the ternary R&D organization within Haier and the peripheral innovation network; it ensures independent development as well as "leveraged" development and ensures the optimal allocation of R&D resources spread over long-, medium-, and short-term innovation projects. The process management system for product development consists of progress assurance system, quality assurance system, and goal optimization system; it implements strict process control for all development projects in accordance with the "OEC" principles. The incentive mechanism for technological innovation is built upon the "competing for post and three workers rotation" scientific labor allocation system, the "combined efficiency" distribution system, and Haier's incentive system; it is an excellent power system for achieving the goals of technological innovation.

2. Internationalization of technological innovation: Haier, going from China to the world

Haier holds three principles in its technological innovation activities. One core principle is internationalization of the objectives of its technological innovation activities, which requires Haier to take a global view, fully and dynamically understand the status quo and development trend of the subject matter, make use of the most advanced scientific research on this subject, and stand on the shoulders of the giants when carrying out innovation activities.

(1) Cooperating with the world's leading high-tech companies and establishing technology alliances

In the process of technological innovation, Haier has always emphasized integrating international innovative resources and improving product competitiveness. Haier has conducted project-based joint research studies with large foreign companies, research institutes, and universities and has set up joint research centers with its partners including Toshiba, Philips, Mize, Lucent, and NetScreen; such cooperation has greatly improved Haier's own ability to meet future changes. For example, when developing color TVs in 1997, Haier established technology alliance with Philips from the Netherlands, and then set up "Haier Digital Technology Research Center Germany" with Mize from Germany for jointly developing digital products (see Table 3).

Table 3 Haier's technology alliances

Partner name	Partner country
Philips	Netherlands
Mize	Germany
Lucent	United States
C-MOLD	United States
NetScreen	United States
Toshiba	Japan

(2) A big leap forward in internationalization of technological innovation: overseas branches

Setting up foreign subsidiaries represents an important step for multinationals in their drive to enter overseas markets. By setting up subsidiaries, multinationals will gradually be able to exploit local advantages in funds, technology, and manpower and provide products for the market according to the local environment. The innovation strategy in Haier's internationalization effort is "difficult ones first"; in other words, first create the Haier brand in the most difficult market and then carry out local manufacturing and local sales. After setting up factories abroad, Haier's secondary goals are to achieve local financing, local mind, and local culture; make use of local manpower and local capital; blend into the lives of local consumers; and develop local brands. Through continued effort, Haier has set up subsidiaries in many countries on the four continents of the world, including Indonesia, Iran, Malaysia, North African countries, the United States, and France.

Haier also has joint development institutes in places such as the EU countries and Japan. In 1999, products designed by overseas design companies were very popular in the domestic market; for example, Haier's "Frost-Free Golden Prince" refrigerators, which were designed by Haier's French design division, were sold out in the domestic market. Currently, there are six most stringent quality standards in the world; Haier has passed all of them. Haier's laboratory has reached the top technology level in the world. It can simulate the harshest environmental conditions of the world to conduct product tests. In addition, it has obtained equivalent licenses from all of the world's six most important certification bodies. In the EU countries, the purchase of Haier's fluoride-free and energy-saving refrigerators can be subsidized by the local government; this advantage has greatly promoted the sales of Haier products in the EU countries. In the United States, Haier products have already met the 2001 United States energy consumption standards.

For the competition in the U.S. market, Haier has adopted a localization strategy; it has established "Haier Design Center" in Los Angeles, "Haier America Trading Company" in New York, and "Haier Production Center" in South Carolina and has integrated the design, production, and sales functions into one business structure. The main purpose of this is to better understand the U.S. market and respond to more rapid market changes. Most of Haier's sales in the United States are not of Haier's original products, but of products specifically designed and produced for the U.S. market. For example, an export item to the United States, Commander in Chief BCD-275—a Haier refrigerator, was especially designed, developed, and manufactured with the feedback from Haier's overseas information station for the Americans on regional characteristics, such as appearance, cooling capacity, usage, and so on. (See Table 4 and Table 5 for Haier's overseas subsidiaries and branches.)

Table 4 Haier's overseas subsidiaries

Subsidiary name	Subsidiary host country
Haier Sha Paul Ltd.	Indonesia
Haier LKG Electric Appliance Co., Ltd.	Philippines
Haier Industrial (ASEAN) Limited	Malaysia
Haier Middle East Co., Ltd.	Iran

(Continued)

Subsidiary name	Subsidiary host country
Haier Design Center (Los Angeles)	United States
Haier America Trading Company (New York)	United States
Haier Production Center (South Carolina)	United States
Haier Product Design Department in France	France
Tokyo Haier Product Design Department	Japan

Table 5 Number of Haier's overseas branches

Category	Number
Manufacturing unit	45
Factory	13
Joint development institute	6
Information center	10
Design center	6

(3) International information sources of technological innovation

What has Haier established worldwide? Haier has established information stations, design centers, and more than 30,000 sales outlets and hired lawyers and consultants from a number of countries worldwide, constituting a global information network, through which Haier can have a timely access to the world's latest technological information as well as market information and fully understand the design trend and legal development in the local markets (see Table 6).

Table 6 Haier's overseas design branches

City/Area	Country
Los Angeles	United States
Silicon Valley	United States
Lyon	France
Amsterdam	Netherlands
Tokyo	Japan
Montreal	Canada

In October 1994, its first overseas product design division, Japan's Tokyo Haier Product Design Department, was formally established. Since then Haier has set up product design divisions in Los Angeles and Silicon Valley, Lyon, Amsterdam, Montreal, and Tokyo, and information centers in Seoul, Sydney, Tokyo, Los Angeles and Silicon Valley, Amsterdam, Lyon, Montreal, and Taipei and Hong Kong (see Table 7).

Table 7 Haier's overseas information centers

City	Country
Seoul	R.O. Korea
Tokyo	Japan
Lyon	France
Los Angeles	United States
Montreal	Canada
Amsterdam	Netherlands
San Francisco	United States
Sydney	Australia
Taipei	China
Hong Kong	China

3. Revelation for Chinese enterprises from Haier's internationalization of technological innovation

Throughout the history of Haier's development, technological innovation has always been the main driving force for growth and expansion. Its growth trajectory has important implications for Chinese enterprises in the process of internationalization of technological innovation.

(1) Haier's internationalization of technological innovation process has gone through the following three main stages: joint venture and technology introduction, establishment of its own technology research centers and development system, and exporting technology or setting up its own R&D branch units overseas. Haier's joint venture in 1984 with Liebherr is the model of technology introduction. During this stage, the

technology for China's home appliance industry was at a level much lower than that of the developed countries. As the learning process extended for technology personnel in production and cooperative R&D, Haier gradually developed its own scientific research teams and moved toward the second tide of innovation. In this stage, it established the R&D center and a sound R&D system within China. Haier's market reach was extended overseas as the company wanted to make use of local talent and get its R&D personnel close to various local markets. Haier's R&D has also begun to expand overseas. We can see that this is an import–absorption–export process. In every stage of the process, internationalized technological strength and talent should be fully utilized.

(2) Setting up a technology monitoring network is the first step in innovation internationalization. Haier has established its own technology monitoring points in multiple countries of the world. These technology monitoring points have multiple functions. The first function is to monitor the trends of its competitors' technological development. The second function is to monitor technological development in its own fields. The third function is to provide information and guidance for localization of technology. A distinctive feature of internationalization of technological innovation by latecomer countries is the setup of technology monitoring network, which is an important means to make up for the deficiency in ability in the transition from imitation to innovation.

(3) Technological innovation capabilities form the foundation for enterprise's internationalization of technological innovation. Haier's internationalization of technological innovation began with the introduction of foreign technology, but Haier took great pains to develop its own capabilities and set mandatory goals to ensure the implementation of technological innovations. Haier has set up R&D centers and has built a sound peripheral R&D system. Such measures have ensured that Haier's technological innovation capabilities have

been improved and have stayed at a high level. Such capabilities can ensure the effective use and absorption of technological information from the overseas technology monitoring sites.

(4) Market-oriented localization of technological innovation is the direction of Haier's development. Haier's technological innovation emerged from importing and absorbing to setting up overseas R&D institutions. Using local talent for the R&D and design of localized products and achieving the localization of international innovation have revolutionary significance for breaking through the inconvenience to management and production brought about by the differences in culture, time and space.

As a company from a developing country, Haier's internationalization of technological innovation is markedly different from that of companies from the developed countries. GE and Motorola have set up their own R&D institutions worldwide; such institutions are generally larger and fewer and usually employ a large number of local talent. However, Haier's R&D units overseas are generally smaller, but in larger numbers, and they usually have strong technology monitoring capabilities. Summing up the experience of Haier's internationalization of technological innovation has important meaning to Chinese enterprises in the process of going international.

Case 3

Holley Group's internationalization of technological innovation

Holley Group is a trans-regional, diversified, export-oriented enterprise group with Holley Group Co., Ltd. Its business activities span measuring instruments, power automation, electronic information, biopharmaceuticals, real estate development, and other areas. Its production bases are distributed in Hangzhou, Chongqing, Guangzhou, Beijing, Shenyang, and other places. It owns manufacturing plants and research institutions in Thailand, the United States, and Canada.

In 2000, Holley Group invested in Silicon Valley, California, the United States. It established a wholly owned enterprise—Holley Holdings (U.S.A.) Co., Ltd., and started to use it as a parent company to invest and develop in North America. The main duties of Holley Holdings (U.S.A.) Co., Ltd. are to follow the group's overall industry development strategy, take capital investment as the means, purchase and acquire high-tech enterprises in North America at low costs, rapidly consolidate the technological resources, formulate the optimal resource allocation with "R&D bases in North America and industrial bases in China," and provide a good platform and effective interface for Holley Group in an effort to enter high-tech fields.

Holley Holdings (U.S.A.) Co., Ltd., since its inception, has acquired a NASDAQ-listed company—United States Pacific System Control Technology Company—which, after asset and business restructuring, has become the window to the United States for Holley's power automation business and has imported advanced power automation technology and products from North America to China for industrial development.

Although this was a notable move by Holley Group, there was another story that was truly captivating and became a hot topic in China's domestic media—the acquisition of Philips's mobile phone R&D division.

1. Chance

Philips Semiconductors is a subsidiary of Royal Philips Electronics Company; it was ranked with Qualcomm and LSI Logic Corporation as the world's three companies selling CDMA chips.

Philips CDMA R&D Division was formerly VLSI Company of the United States. In 1999, Philips acquired it for $120 million and invested another $80 million later to maintain R&D. In other words, Philips invested more than $200 million in the CDMA R&D Division. However, the global slump of the IT industry made "unloading baggage" a popular choice for international mobile phone companies. In early May 2001, with no good business prospect in sight after significant investments as mentioned above, Philips decided to suspend further capital input and started to consider selling the division. Via

Philips CDMA China Sales Advisor, Ni Weizu, who enjoyed a close relationship with Holley, this message was instantly passed over to Wang Licheng, who was at that time trying to reach Philips for cooperation in making mobile phone components. Holley immediately contacted Royal Philips Electronics Company.

The acquisition achieved its initial success within three months. However, for a company with a Chinese background, it took an unusually difficult process to win the trust and acceptance of Philips that had a long history. They were asked to hand over a guarantee of $2 million for entering the negotiation; this requirement made the employees of Holley feel that Philips displayed a harsh attitude toward a Chinese company.

Before then, all the attempts made by companies from Taiwan of China to acquire a U.S. company had failed in the end due to the ultimate differences in culture and in management ideas, so that many buyers ended up getting only some static stocks—equipment, technical materials, and products—and technology talent who formed the core of the company left. However, there was a huge pressure for operating capital, millions of USD of upfront expenses, and tens of millions of USD of yearly input for maintaining R&D in CDMA technology and for operating expenses, thus leading to more pressure on Holley.

In September 2001, Holley Holdings (U.S.A.) Co., Ltd., after months of arduous negotiations, officially acquired the CDMA mobile communications division located in San Jose originally owned by Philips and won all the intellectual property rights, R&D outcome, R&D equipment, R&D tools, and a large number of experienced R&D personnel owned by Philips in the field of CDMA wireless communications. Then, on this foundation, it set up Holley Communications Group, Inc., which specialized in the design and development of mobile phone chip software and mobile phone's complete technology solutions, went through Zhejiang Holley

Communications Technology Co., Ltd. to the Chinese market, and provided core chips and complete technology solutions to China's mobile phone manufacturers, thereby becoming China's first enterprise in possession of the complete set of core technologies in the IT industry.

After the successful acquisition, Holley openly declared that Holley Communications Group is "an American company with Chinese investment background and Chinese market background," and not the U.S. subsidiary of a Chinese company in the traditional sense. Both the CEO and CTO of the company are American citizens, all management and operations are done in accordance with U.S. practice, and the company announced a four-year option plan immediately after the acquisition. Such measures removed the worries of the U.S. employees about the company "operating at a low level and doing things in the Chinese way," and almost all of the overseas employees extended their employment contracts. On October 16, 2001, Holley announced that the acquisition had succeeded, the CDMA IS-95 chip being worked on would be released in the first quarter of the following year, and GSM/CDMA dual-mode control chip would enter the market in the second half of the following year.

2. Competition and cooperation with Qualcomm

Holley realised that Qualcomm's absolute dominance in CDMA wireless system chip and system software would not be shaken even by the full cooperation between Holley and Philips. It was reported that Qualcomm had already received $600 million in technology cooperation fees, which was less than 20% of the overall fees for complete technology transfer.

In fact, Holley is only responsible for chip design, while production and sales still go through Philips, and the profit will then be split between the two parties. When first getting into the CDMA chip market, Philips won the right to use the patent for the CDMA chip and for the production license. However, Holley, only through acquisitions, obtained the former. It claimed that the CDMA IS-95 mobile phone technology solution being launched would lower the cost for a CDMA phone by at least $20. Based on the 10-million capacity for the first phase of the China Unicom CDMA network, it was estimated that this new mobile phone technology solution would save $200 million for domestic cell phone makers. Holley claimed the following reasons for this:

- Low purchase price;
- Sharply slashing operational costs;
- "Boeing's aircraft manufacturing" strategy: hold the integration technology in its own hands and outsource all the component technologies. Previously, Philips worked in all the related projects independently, and the total costs for manpower, time, and financing were astonishing.

A saying has been circulating in the industry: once domestic cell phone makers stop buying Qualcomm chips and start to buy from Holley, Qualcomm would charge such "violators" 6.5% royalties for using their patents. For this, Holley said that they were not worried because, "Even so, we are still very confident. What we will do is to make sure that, even after paying 6.5% royalties to Qualcomm for using their patents, it will still be cost-effective to use our chips."

The acquisition of Philips CDMA not only has enabled Holley to bypass Qualcomm's technology monopoly, but, at the same time, Holley has itself become a holder of core monopolistic technologies as well. A notable commonality between Holley and Qualcomm is that if a domestic manufacturer is to buy Holley chips, it should pay technology royalties to the latter.

However, some people believe that, in the production of CDMA chips, Holley should cooperate with Qualcomm. Holley can help Qualcomm make a bigger pie of the CDMA market, and

Qualcomm should lower the barrier of entry, in order to increase the competitiveness of the entire CDMA industry relative to the GSM; the benefits from cooperation can be a lot more than those from competition.

3. Reflection

Foreign business giants have resorted to the advantages of their technologies and strong financial resources and have dominated the entire global market. In contrast, even after ups and downs, China's communications enterprises are often still operating at the bottom of the whole industrial chain and playing the menial roles such as assemblers and porters in the seemingly huge end-product manufacturing area. The Chinese enterprises are in a similar situation in the fields such as software and large scale integrated circuit (LSIC), where China does not control the core technologies.

After China's entry into the WTO, for Chinese domestic IT enterprises that always prefer brand marketing to technological innovation, how to get a foothold in the international market? One model is probably more worthy of adopting than independent R&D which takes a very long time: to acquire the most central core technologies overseas via effective M&As in the overseas capital markets and enter the global market at top speed. Holley's acquisition is a reflection of this thinking.

Holley's acquisition has also touched on some deeper-level questions about overseas investment by China's domestic enterprises. Some economists noted that international investment is actually the process of resource allocation globally; the core ideas of investment from now on will be focused on the optimal allocation and utilization of technology resources, market resources, human resources, and capital resources.

Simplistic investment thinking, such as “sell my own products,” is now questionable and actual situations vary. While the investment in South Africa’s electronic appliances by Shanghai Guangdian Group is for the large potential market for televisions and washing machines, the acquisition of Philips’s R&D institution by Holley Group is for the talent, capital, and technology advantages of the United States—the market advantage is still in China.

Case 4

Huawei's internationalization of technological innovation

Since the second half of the 20th century, with economic globalization and rapid development in technology, competition among enterprises has become more and more globalized and geographically dispersed international technological innovation activities have become the main driving factor for enterprises to win in the worldwide competition.

As early as in 1986, the number of overseas laboratories set up by companies of the Netherlands and Switzerland exceeded that of their domestic laboratories (Pearce & Singh, 1992). During the period from 1983 to 1993, the growth rate of overseas R&D investment made by the U.S. companies was three times that of their domestic R&D investment. During the same period, the percentage of their overseas R&D spending in the total R&D investment grew from 9% to 15%. The amount of overseas R&D activities carried out by the European companies was a third of that of their domestic R&D activities. Since the mid-1990s, the R&D investments by the world's 50 largest multinationals in the three main regions—the United States, Japan, and Europe—have accounted for 33%, 57%, and 42% respectively, of their total domestic R&D investments (Gassmann & Von Zedtwitz, 1999). Internationalization of technological innovation has become the necessary condition and common form for companies to improve their technological capabilities and effectively participate in international competition.

Now, China has acceded to the WTO; this means that the domestic market and the international market are increasingly integrated, that domestic supply and demand will no longer be the primary deciding factor for the rise and fall of an enterprise, and that the national economic development and the fate of each enterprise will be rearranged on the larger stage of the world economy. Only by hastening the upgrade of strategic position for technological innovation and speeding up the implementation of the strategy for internationalization of such innovation, can Chinese enterprises catch up with the pace of worldwide technological development and attain strategic successes in both national-level and enterprise-level technological innovation.

Communications equipment manufacturing is one of the fastest growing industries in China, and the reason for the success lies in that the Chinese enterprises have closely followed the technology development trend in worldwide telecommunications equipment manufacturing industry, made full use of worldwide technology resources, have leapt-forward in development through international technological innovation, and thereby gained a place in the fierce competition of that worldwide industry. Shenzhen Huawei Technologies Co., Ltd. is presently China's largest telecommunications equipment R&D and manufacturing company, and China's other companies can gain good insights from Huawei's experience in internationalization of technological innovation.

1. Company profile

Shenzhen Huawei Technologies Co., Ltd. was established in 1988. The company's independently developed products cover the main telecommunications fields, including exchange, access, transmission, mobile communications, intelligent network, support network, ATM, access server, router, Ethernet switch, videoconferencing, etc. The company has formed its own core technology system; provides a full range of solutions in fixed network, mobile network, and data communications network; and is a worldwide leader in the fields of optical network, mobile communications, and broadband technologies.

In a short span of a few years, Huawei has come to occupy the top position among China's telecommunications equipment manufacturers and has become a powerful competitor for many of the international big-name telecommunications companies in the Chinese market. The main reason is that Huawei has always treated technology development as its topmost priority, formed its own core technology system, and continually developed new products that meet the needs of the market. Today, more than ten products of Huawei have been ranked among the advanced or leading products of the world's telecommunications industry, and Huawei has gained a status in technology equal to that of the world's other well-known companies and has gained a foothold and achieved development in the telecommunications equipment market that has seen an increasing speed of elimination and is fiercely competitive.

Huawei has a very clear strategic goal to become a world-class and leading telecommunications equipment provider. To this end, Huawei sees technological innovation as the foundation for maintaining its place and for further development and sees internationalization of technological innovation as the company's technological innovation strategy.

2. Huawei's strategy for internationalization of technological innovation

Seeing Ericsson, Siemens, and the likes of multinational communications giants, Huawei knew that they started late and that their strength was lacking. Therefore, Huawei at its beginning stage developed a two-point development strategy: first, do not carry out a full-range catch-up and instead invest all the strength in developing core network technology, take the road of industrial specialization, and focus on telecommunications equipment; second, adhere to the road of internationalization of technological innovation. By directly establishing overseas R&D institutions, carrying out overseas M&As, forging international technology alliances, and by other means, the company extended its innovative activities abroad. Based on the globalized acquisition of innovative ideas, internationalized innovative resources, and networked organizational form, the company shaped its own globalized R&D system. It formed an enormous knowledge transaction network operating inside the company and linking the company and outside organizations and constructed its own technology knowledge structure with its external knowledge and information sources.

(1) Strategic goals

The internationalization strategy of Huawei's technological innovation is to drive forward comprehensive internationalization of the company's production and operations with internationalized technological innovation. Two reviews of the process of internationalization of technological innovation by the world's major multinationals revealed that the process actually begins with internationalized production and operations, then goes through a stage of internationalized technological innovation, and eventually reaches the stage where internationalized technological activities support the geographic expansion of production and marketing (Cantwell, 1995; Cantwell & Piscitello, 2000). The initial motive for multinationals to internationalize technological innovation is to carry out technology transfer. After production and other business operations have become internationalized, the company sets up overseas R&D institutions so as to make maximum use of the commercial potential by available technology and help overseas production units to make adjustments to their products and technological processes in order to adapt to the local production conditions and customer preferences. On such basis and for long-term strategic considerations, multinationals gradually build up their global innovation networks so as to maintain their competitiveness, make use of various technology resources of different countries, and lower the costs and risks involved in new product development. Huawei believes that, as a relatively new enterprise in the industry, it must make full use of the advantages of being a latecomer, learn from the world's well-known companies of their technological innovation experience, embark on the path of internationalization of technological innovation right from the beginning, and drive forward the overall growth of the company accompanied by rapid increase in technological capabilities.

In the process of implementing internationalization of technological innovation, Huawei's strategic direction is very specific (see Figure 1), that is, to extensively absorb the results of the latest research worldwide in the field of electronic information, to learn with an open mind from outstanding companies at home and abroad, and to develop, on the basis of independent work, a leading system of core technologies openly and cooperatively. Through internationalized technological innovation, Huawei aims to achieve three objectives:

1) Establish information windows, track as well as obtain technology from host countries and competitors, gain from technology spillovers of the local R&D, and use them as supplementary sources of specialized technology.

2) Get close to knowledge centers, make use of S&T talent and

research environment of the host country, lower R&D costs, and improve efficiency of technological innovation.

3) Respond to the special needs of various customers worldwide and to the local production conditions, quickly react to the differences among the various markets, localize technology so as to support Huawei's production enterprises in the host country, and effectively allocate its capabilities.

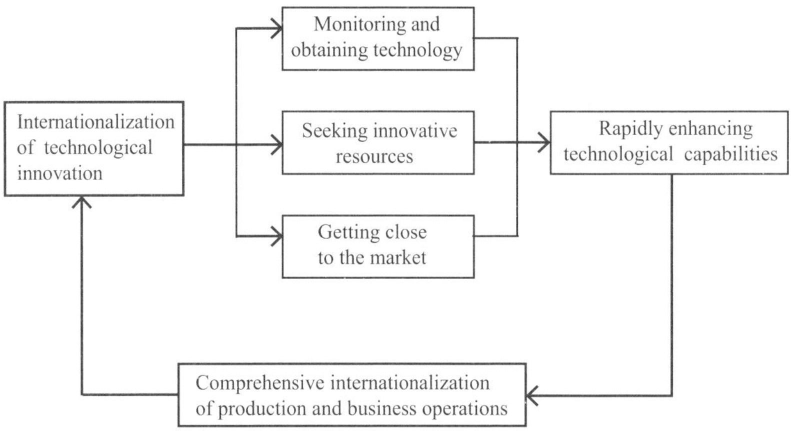

Figure 1 Huawei's strategy for internationalization of technological innovation

(2) Strategic development approach

To achieve the established strategic objectives, Huawei devised a four-stage development strategy.

The first stage is to establish technology alliances with foreign companies, engage in technology learning in the cooperative process, and improve the ability of technological innovation.

Huawei's initial process of internationalization of technological innovation is the one featuring "coming in." While independently developing proprietary products, Huawei actively engaged in cooperation with domestic and foreign companies, universities, and research institutes and used such cooperation as an important source of innovative ideas and the main approach to enhancing its technological strength.

Huawei has established joint laboratories with TI, Motorola, IBM, Intel, Lucent, Altera, Sun, and other world-class companies one by one; set up long-term, friendly, open, and win-win cooperative relationships; extensively engaged in cooperation with both technology and marketing; and achieved internationalization in both technology development and cooperation.

The joint laboratory on digital signal processing established by Huawei together with TI has been committed to jointly developing DSP products; Huawei–Lucent joint laboratory engages in comprehensive cooperation with microelectronics, optoelectronics, and some other areas. However, the main tasks for the Huawei–Sun joint laboratory are to organize applied technology training and exchanges for Sun products (systems and software/hardware) and carry out cooperative R&D based on the needs of product R&D. Huawei and Sun also planned to use the laboratory for applied research projects such as business support platform, load-balanced high-availability clustering technology, distributed databases, JAVA application in the telecommunications field, and so on.

By carrying out cooperative technology development with technologically strong multinationals, Huawei can quickly obtain the required technology and acquire expertise in research management. Successful technology alliances helped rapidly improve Huawei's technological capabilities; with its strength growing stronger, Huawei has gradually become the backbone of China's telecommunications equipment manufacturing industry.

The second stage is to keep track of worldwide technological developments, get close to the world's technology centers of excellence abroad, and set up overseas R&D institutions.

As Huawei gained further strength, it started to go out of China and set up overseas R&D institutions. At this stage, Huawei's market was still concentrated in China and Huawei competed with other domestic companies and the subsidiaries of overseas multinationals, as such was the main feature of Huawei's business at the time. Therefore, its objectives for setting up overseas R&D institutions at this stage were to directly learn and monitor worldwide technology developments and to get close to the world's technology centers of excellence, so as to better benefit from the spillover effect of foreign R&D activities.

Studies done by Cantwell and Hodson (1991), Hakanson (1992), and Pearce and Singh (1992) showed that for the R&D activities of the technology monitoring or acquisition type, the distribution is positively correlated with the overall distribution of the world's innovative activities. However, R&D activities of the basic research type should be carried out in cooperation with the local universities and research institutes overseas, and thus, it is very important to be close to the overseas technology centers of excellence and choosing a location close to good universities or research bases in the same field of research is the key factor to ensure the success of the planned technological innovation activities.

For selecting locations for its overseas R&D units, Huawei attaches great importance to scientific methods (see Figure 2). The United States is the center of S&T development in modern times, and Silicon Valley is a world famous high-tech base. For monitoring and gaining access to the latest developments in optical network technology and other telecommunications products, Huawei has set up Huawei USA (Silicon Valley) and Dallas Institute of the United States, which are mainly engaged in optical network technology development and in the research of other telecommunications products. Sweden is the center of mobile telecommunications technology such as Europe's GSM and

WCDMA, and Huawei Sweden Institute is responsible for monitoring the development of Europe's GSM and WCDMA technologies and for carrying out research in mobile telecommunications technology.

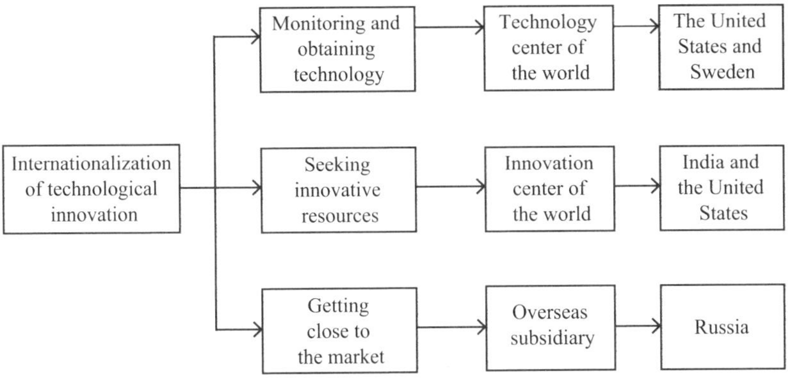

Figure 2 Location selection for Huawei's overseas R&D units

For R&D activities of the resource seeking type, location selection of overseas R&D units is mainly influenced by the supply of host country's S&T talent, technical facilities, and the level of protection provided by the environment (Kumar, 2001). CMM was developed in 1991 upon the request of the federal government of the United States. It is a model for evaluating the capabilities of a software contractor and for helping it to improve on quality, and also the world's most popular and most practical standard for software production process as well as the standard for certifying the maturity of a software enterprise. India has the best CMM environment in the world. In order to learn from India's open software development process and quality control methods as well as to make full use of India's software research resources, Huawei—in June 1999—set up the India Research Institute in Bengaluru, also known as India's "Silicon Valley," Huawei recruited local talent and made full use of the local research environment and other resources.

The third stage is to provide support for international market development and set up overseas R&D institutions that innovate based on the local market conditions.

With continued improvement in technological capabilities and the gradual growth of its domestic market, Huawei began to turn to the international market. International operations have become the main objective of Huawei's business strategy. With continued development of its international market, Huawei has discovered that even in the telecommunications equipment industry that has a relatively high degree of standardization, there are still differences among various regions in market conditions and customer preferences. To effectively expand its overseas markets, Huawei started to set up overseas R&D units for various market conditions overseas.

A study by Kumar (2001) showed that R&D activities of the production support or market-oriented type are mainly for making adjustment to technologies and products from the home country, and thus, adaptation to the market conditions of the host country and proximity to the existing production subsidiary of the parent company and the main customers are very important. The establishment of Huawei Russia Research Institute is undoubtedly a success story. In order to seize Russia's wireless communications market, Huawei relied on Russian Beto-Huawei joint venture, set up Russia Research Institute that specializes in localizing Huawei's technologies, and has gained a large share of the local market there.

The fourth stage is to carry out diversified, multisite, and global technological innovation activities.

Huawei has become China's biggest telecommunications equipment manufacturer and has set up branches or joint ventures in more than 40 countries. Huawei's technological innovation has reached a stage of diversified, multisite, and global activities. Technology alliance, overseas R&D institutions, and domestic R&D institutions for different regions together form Huawei's global R&D network (see Figure 3). Through this R&D network, Huawei achieved a network-based collaboration model for technological innovation.

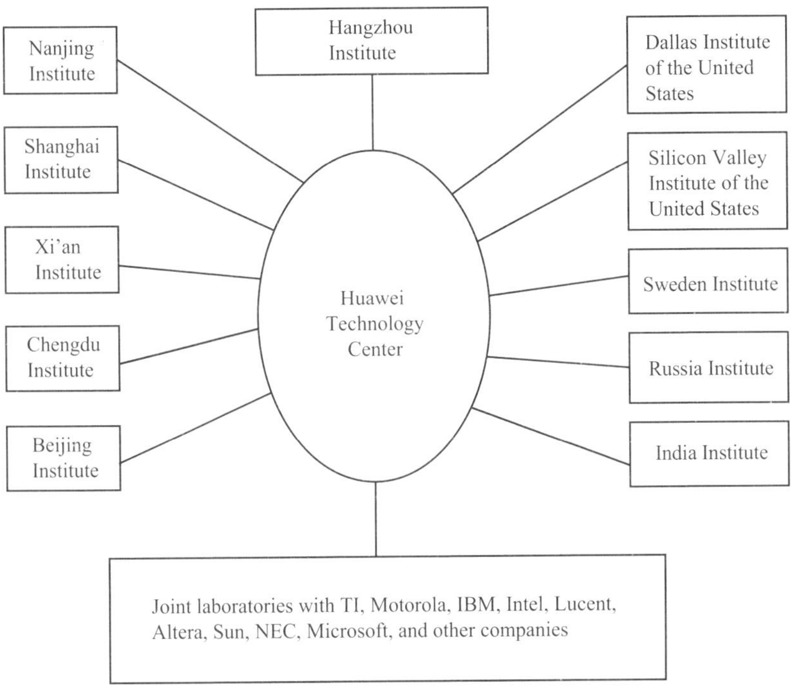

Figure 3 Huawei's global R&D network

The R&D of CDMA technology is an example of success with Huawei's network-based collaborative innovation model. To China's telecommunications manufacturing industry, the development of CDMA can become a rare opportunity to fight a decisive winning battle in the field of mobile communications. This is because of the fact that, in this area, in terms of both technology and market, all the domestic and overseas companies are at the same starting point and operate in the same environment. Since 1995, Huawei has been committed to the R&D of CDMA-related technologies. Huawei's CDMA products are the fruits of its network-based international collaborative innovation. RF (Radio Frequency) technology research is done in Huawei's institutions in Russia and Sweden; software-related technologies are developed by Huawei's institutions in India and the United States, which have cooperated with the local research institutes, fully absorbed local talent and technology, and achieved internationalized resource utilization. On the domestic front, each of Huawei's research institutes in Beijing, Shanghai, Nanjing, and so on has its own specializations, and Huawei has entered into extensive cooperative research relationships with Beijing University of Posts and Telecommunications, Tsinghua University, Northeastern University, University of Science and Technology of China, University of Electronic Science and Technology of China, Xidian University, Xi'an Jiaotong University and Fudan University in key mobile telecommunications system technology fields, including CDMA, and has actively participated in standards development and research. The various R&D institutions have achieved information sharing on the basis of task division and collaboration on the overall integration of CDMA technologies.

On October 12, 2001, Huawei's CDMA 1X base station products were officially released, marking the point when China's mobile telecommunications manufacturing industry, in the CDMA field, started to run in sync with the world's leading companies.

3. Organizational structure and resource allocation of Huawei's internationalized technological innovation system

In the process of Huawei's internationalization of technological innovation, each development stage is not clear-cut and absolutely distinct; instead, it is a stage when the company, on the basis of the actual situation, establishes a principal direction, and, in accordance with the strategy and form of the innovation internationalization, makes proper adjustments to the organizational processes and resource allocation.

The form of Huawei's internationalization of technological innovation, during various development stages, has gone through the transition from being centralized to being network-based. With the changes in the form of internationalized innovation, adjustments have been made to the organizational structure and resource allocation of Huawei's internationalized innovation system.

During the initial stage of Huawei's internationalization of technological innovation, innovative activities were mainly concentrated on the technology center of the company's headquarters, and the R&D branches were mainly responsible for gathering information about technological innovation and for some small ancillary projects. Accordingly, decisions about setting goals for overseas R&D, planning overseas R&D projects, and recruiting overseas R&D personnel were made at the company's headquarters. The R&D branches were coordinated by the company's headquarters and its R&D center, and there were minimal interactions among the branches themselves. The entire internationalized R&D system displayed a star-shaped structure (see Figure 4). For resource allocation, most key strategic assets were concentrated in the company's headquarters. Information about technological innovation flowed unidirectionally from the branch units to the technology center of the company's headquarters, and information about decision-making flowed unidirectionally from the company's headquarters to the branch units.

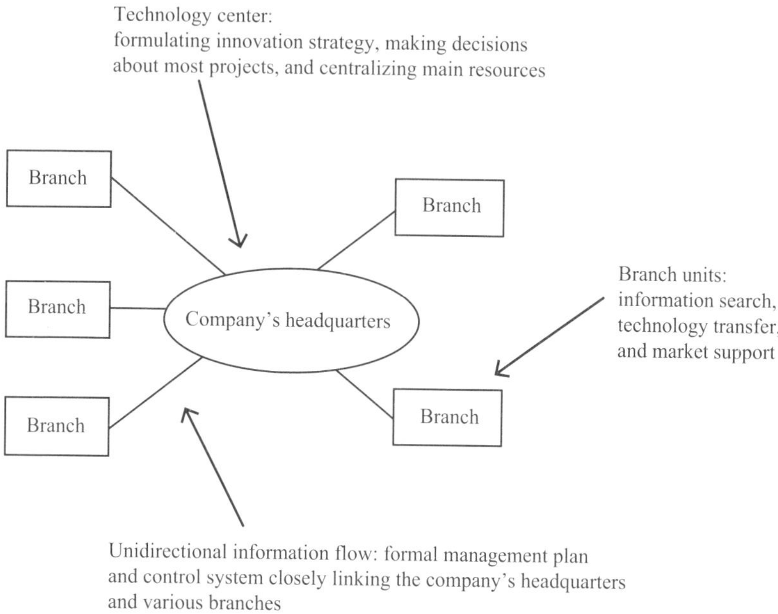

Figure 4 Star-shaped structure of Huawei's internationalized innovation system

During the early period when the company was weak, concentrating on the key resources such as knowledge and technology under the star-shaped structure could help Huawei focus on the resources for technological innovation, take advantage of the scale effect, and rapidly develop new products and processes. However, as the company's market expanded, its technology system became more complex. When implementing technological innovation at a higher level, the downside of this organizational structure showed up. First, when facing different market conditions, the R&D center at the company's headquarters lacked knowledge about overseas market conditions and the actual needs, and thus, the company could respond fast enough to different market conditions. Overseas R&D branches—because they simply implemented R&D projects and had limited resources and lacked the opportunity to learn—did not have the ability to respond to the changes in market conditions adequately. Second, when carrying out innovation of complex products or developing multiple technological products, resources from multiple parties would be needed and collaboration involving multiple parties would be necessary; however, R&D branches—because of their limited capabilities—were not able to acquire adequate resources, and the weak cooperation within the system made the system unable to adapt to the needs of such innovation.

As Huawei's technology system became more complex and its international market further expanded, Huawei's internationalized innovation system began to evolve into an integrated network (see Figure 5).

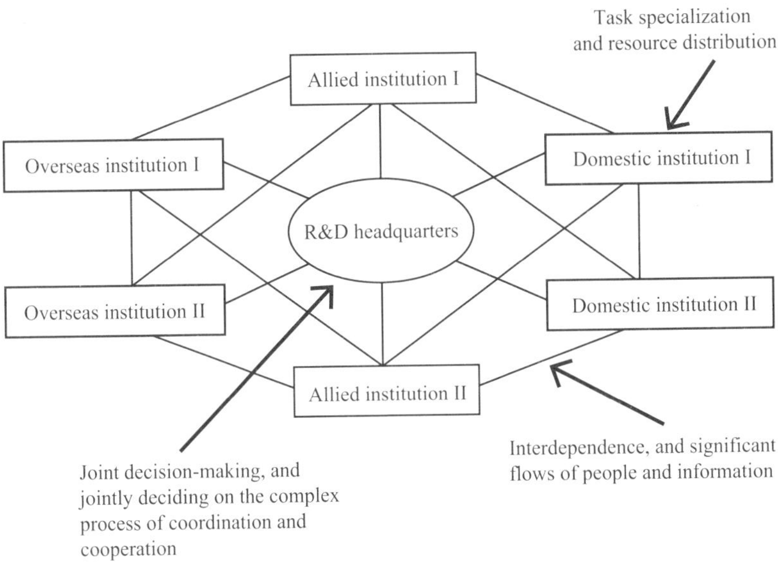

Figure 5 Huawei's network-based R&D system

In the process of building an integrated innovation network, the first step Huawei took was to distribute assets and strengthen various R&D branches. The R&D branches were no longer seen as the units for technology search and information acquisition, but as the units for technology development. This measure greatly strengthened the capabilities of the company's R&D branches.

The second step in Huawei's effort to build an integrated R&D network was to define the role of each unit and achieve specialization. Each R&D branch was in a unique local environment and enjoyed a unique combination of R&D resources. In accordance with the actual situation of each branch, the company assigned specialized tasks to it; different branches had different specializations. Meanwhile, depending on the environment resources of each R&D branch, the company defined its place in the company's overall R&D system. Those R&D branches with a local environment of great strategic importance, rich resources, and strong capabilities would be identified as the strategic leaders of the overall R&D system, and their main tasks would include discovering opportunities and participating in formulating the company's R&D strategy. Those R&D branches with a local environment of limited strategic importance, rich resources, and very strong capabilities would be identified as doer branches; their tasks would be to carry out R&D activities in their areas of specialization.

The third step in Huawei's effort to build an integrated R&D network was to change the management approach. With regard to decision-making, R&D branches were given more autonomy. The company's headquarters only made important decisions involving overall strategic projects. Overseas R&D units had the power to make decisions on the planning of common projects, recruitment, and use of funds. The management of the company's headquarters supervised the operations of various R&D units through annual performance appraisal, so as to lower the costs and improve efficiency; with regard to exchange and communication, more flexible methods of coordination were adopted. In addition to formal planning and control, parallel and informal exchange and communication were also promoted.

Through these steps, Huawei has formed its own integrated R&D network. In this network structure, decision-making is centralized and relatively decentralized and resources are distributed in accordance with the strategic position of each unit within the overall organization and its objective; various units are both interdependent and specialized, and at the same time, significant flows of resources, manpower, and information happen among the various units.

Bartlett and Ghoshal (1990), Brockhoff (1998), Gassmann and Von Zedtwitz (1999), and other overseas scholars, through research, have concluded that the organizational structure of an innovation network is built upon the principle of specialization and interdependency of various R&D units; that such interdependence in resource and responsibility breaks the contradiction between local interests and global interests; and that, as a result, companies can acquire important knowledge from the market, the world's technology centers of excellence, and even the competitors by sharing resources, knowledge, and opportunities, and greatly improve the efficiency of enterprise innovation through cooperation among the various R&D units. Globally interconnected innovation network has become the trend in the development of organizational structure for enterprise innovation. Huawei's network-based integrated innovation system will undoubtedly be a powerful driving force for the rapid development of its technology.

4. Revelation of Huawei's internationalization of technological innovation

Compared with the overall situation among the Chinese companies, Huawei's internationalization of technological innovation has reached a very high level. The path Huawei has taken and the experience it has gained provide good lessons for Chinese companies in pursuing a strategy for internationalizing technological innovation.

(1) Driving internationalization of production and other operations through internationalization of technological innovation

Huawei started late in global competition, had limited strength, and was at a low level of internationalized production and other operations. In spite of such a situation, Huawei has successfully walked along a path from internationalization of technological innovation to market internationalization. On the one hand, through internationalization of technological innovation, the company can rapidly improve its technological capabilities and, on the other hand, it can stay abreast of the technology trend and market trend of the world's telecommunications industry, thus leading to active participation in international competition.

(2) Making use of overseas technology talent, monitoring, and obtaining technologies of the host country and the competitors, taking full advantage of the world's technological innovation resources, and independently providing innovative ideas for core technologies

The objectives pursued by multinationals when carrying out internationalization of technological innovation can be categorized into three types: (i) to meet the special needs of various customers worldwide, adapt to the local production conditions, achieve technology localization, support the manufacturing enterprises of the company in the host country, and effectively allocate the company's capabilities; (ii) to gain proximity to knowledge centers, set up a window of information, monitor and obtain technologies of the host country and competitors, and benefit from the technology spillovers of the local R&D as well as use such spillovers as a different supplementary source of specialized technology; and (iii) to make use of the technology talent and research environment of the host country, lower R&D costs, increase the efficiency of technological innovation, and raise the overall technological capabilities of the company. Since its weakest area for international competition was in its technological capabilities, Huawei chose the strategy for internationalization of technological innovation by making use of overseas technology talent as well as monitoring and obtaining technologies of the host country and the competitors.

(3) Aligning the organizational form of the innovation system with the strategy

As the strategic emphasis changed when its internationalization of technological innovation went through different stages, Huawei made adjustments to the organizational processes and resource allocation, so as to match them to the strategic objectives of the time and ensure the efficiency of technological innovation. In the early stage when innovation activities were centralized, Huawei adopted a star-shaped organizational structure so as to take full advantage of the scale-of-economy effect in technological innovation. Since Huawei entered the period of comprehensive development, with the unprecedented increase in the size and complexity of its technological innovation activities, an integrated and network-based innovation system has helped Huawei improve its learning capabilities, make full use of the world's technological resources, and enhance the efficiency of its technological innovation.

Case 5

GE's internationalization of technological innovation: experience and what it teaches us

Technology monopoly has always been a key factor in the growth and development of multinationals. Meanwhile, in almost all the Western countries, multinationals are the backbone of technological innovation and are also the earliest creators and monopolizers of the vast majority of industrial technological innovations (Li, 2002). According to statistics, in some developed countries, R&D expenditures by multinationals account for around 70% of the overall national industrial R&D expenditures, and in some countries, this figure can reach above 80%. At the macro-level R&D investment, many U.S. companies spend a lot of money overseas for R&D activities. According to *Globalizing of Industrial Research and Development* published by the U. S. Department of Commerce, between 1987 and 1993, the total investment abroad in R&D by the U.S. companies increased from \$5.2 billion to \$9.8 billion. Such investments are not forever concentrated in the developed countries only; the growth rate of such investments in R&D done in the developing countries is actually quite high. From 1989 to 1993, the total investment in R&D done in Ireland by the U.S. companies increased from $134 million to $669 million; in Brazil, from $940,000 to $220 million; and in Spain, from $115 million to $320 million. However, in Singapore, the total investment by the U.S. companies has jumped from $5 million to $312 million, an astonishing pace of growth (Wang, 1997). At the micro level, many well-known multinationals have set up institutions abroad and have started R&D work there. GE, founded by the great inventor Thomas Alva Edison, is the earliest U.S. company to conduct R&D activities (Liu, 1998); the path it took for internationalizing R&D, the lessons it took, and the experience it gained are noteworthy.

1. Historical path of GE's internationalization of technological innovation

The General Electric Company, GE for short, is a time-honored "century-old enterprise." GE was born in 1892, but its predecessor can be traced all the way back to 1878 and to the famous inventor Edison's Electric Light Company. In 1889, Edison Electric Light Company merged with several companies and formed Edison General Electric Company. In 1892, Edison General Electric Company and Thomson-Houston merged and formed the General Electric Company.

Technological innovation at the General Electric Company goes back the furthest to 1876 and to Edison's Menlo Park laboratory. In 1990, based on that laboratory, GE established Industrial Research Laboratory. The establishment of that laboratory has had great significance in the history of technological and industrial development in the United States. It marked the transition from individual-based S&T research to collective research and signified the emergence of industrial laboratories (Liu, 2001). Although for a long time after the presence of industrial laboratories, enterprise R&D activities were still concentrated in company laboratories and did not cross national borders and the birth of industrial laboratory provided the foundation for internationalization of technological innovation. Prior to this, R&D activities were the endeavors of individual scientists working alone; however, the nature in industrial research of the pursuit of efficiency and effectiveness demanded the transition from individual-based research to collective research. Such transition laid the foundation for the earliest form of internationalization of technological innovation—recruiting technical experts from other countries.

Industrial laboratories in the United States are created based on Germany's system of S&T as the model. In the late 19th century, Germany's system of education, science, and technology was the most developed and most successful in the world. From 1870 to 1915, a large number of U.S. scientists studied and received their PhDs in Germany; they witnessed the reform of Germany's system of scientific research and the effectiveness of the reformed system. These American scholars, educated in Germany, were the main promoters of the transformative changes in scientific research in the United States.

It is worth mentioning that the idea of creating GE Industrial Research Laboratory was first proposed by the company's then Chief Engineer Charles Proteus Steinmetz. Steinmetz was a German in exile in the United States and was a staunch socialist. Before becoming an employee of GE, he was working in a company started and owned by another German in exile in the United States, Rudolf Eickemeyer. In 1892, GE bought that company and recruited both Steinmetz and Eickemeyer. The reason was that the two had made major inventions in electrified railway. Under the social environment at the time, it was a bold move for GE to recruit technical experts from Germans in exile. In a sense, the emergence of industrial research laboratories was the result of international talent recruiting. It was also an early instance of internationalization of technological innovation through M&As.

Even more noteworthy is the fact that the reasons Steinmetz gave in his effort to persuade high-level management of the company to set up the industrial research laboratory were also based on a view on the background of internationalized technological innovation. At the time, GE's lighting technology was mainly high-vacuum carbon-filament incandescent lamp, while the patent of the ceramic incandescent lamp invented by Nernst was sold to Westinghouse; then, at that time, a German professor developed mercury-vapor lamp. Westinghouse gave the professor strong support and was preparing for production. Steinmetz wrote about this situation as a somewhat "alarming" report to the top management, claiming that, in order to prevent others from becoming the front-runner in developing mercury-vapor lamp, GE must speed up its own R&D. Finally, the top management approved the report and a new era began.

Even after a hundred years, the approach that GE adopted for internationalizing technological innovation, which inherited its past style, is mainly focused on recruiting world-class technical talent. However, it should be pointed out that, overall, the degree of its internationalization of technological innovation is very low. Even in recruiting talent, most of the recruited are still domestic talent.

For this period, Irving Langmuir is someone we have to mention. Langmuir is an American scholar who studied at the University of Göttingen in Germany with the famous scholar Nernst. In 1906, he received his PhD degree and went back to the United States. Langmuir made achievements across multiple fields of research. He wrote more than 200 scientific papers, held 63 patents, and won the Nobel Prize. In addition, the first two directors of the laboratory—Whitney and Coolidge—both had studied in Germany. At that time, Germany was extremely advanced in S&T and people with learning experiences there could help the company keep up with the forefront of technological innovation and gain more contact with international institutions in technology R&D. Under the leadership of these people, most of the people in the laboratory were local talent, but there were also some from the University of Munich and the University of Göttingen in Europe. Langmuir's case reflected GM's thinking and strategy of going through a variety of channels to make use of international intellectual talent and gain access to international technological information.

Promoting the flow of international technological information was another characteristic of GE during that time. Whitney and others gained the most advanced information internationally through hosting seminars and lectures. The world's renowned scientists also came to the GE electric laboratory for giving lectures, including Sir W. Bragg (the pioneer of X-ray diffraction), Niels Bohr, and others.

In addition, there were also a small number of cross-border cooperation agreements. Whitney and Coolidge went to Europe separately—the former in 1906 and the latter in 1908—to study the new progress of incandescent light research. GM and European lamp manufacturers reached an agreement of cooperation to exchange technical information (Liu, 2001).

In 1973, a GE scientist was awarded another Nobel Prize. The winner, Ivar Giaever, was a Norwegian and was in charge of the company's R&D work and a backbone researcher, which showed the willingness of GE to accommodate international talent. It also showed the degree of internationalization of its cutting-edge technology researchers.

Figure 1 illustrates GE's internationalization of technological innovation from its founding to the early 1990s.

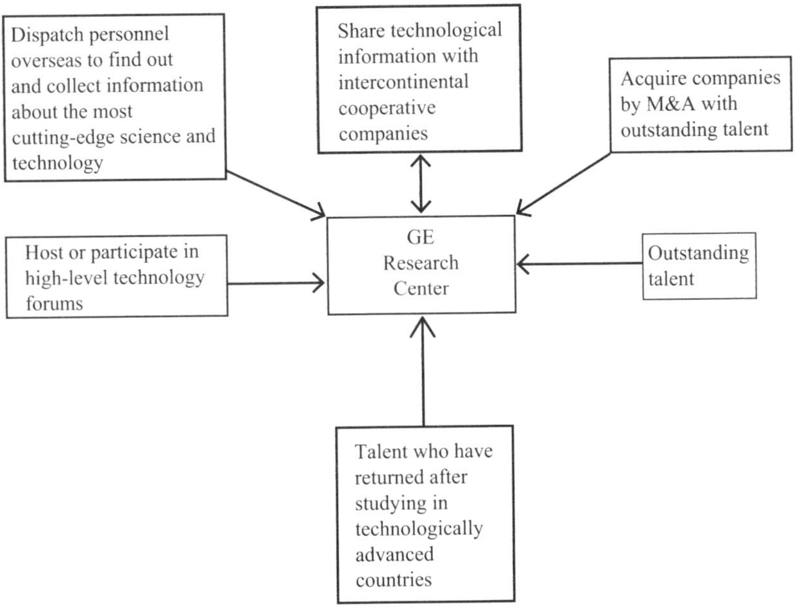

Figure 1 GE's international R&D network (prior to the 1990s)

In 1965, GE merged the Industrial Research Laboratory and the Advanced Technology Laboratory (founded in 1895) to form the Research and Development Center. The center has been the main place for GE's R&D. For internationalization, it was mainly accomplished through promoting information flow at the frontier of international technology research and recruiting world-class talent to work at the center. The center is a global organization, and there are more than 1,600 employees (Liu, 2001) who are from different countries. Among the employees are more than 100 leading technology elites from around the globe, including a number of Chinese scientists (Jiang, 1996).

2. Internationalization process of GE's technological innovation

Until the end of the 1970s, the degree of GM's internationalization of technological innovation had remained at the level of the international flow of personnel and information. This was the result of the objective laws of economics and the reality of the international political economy at the time. Toward the end of the 1970s, Reginald Jones was facing retirement. Psychological research shows that for the choice of successors, by human nature, corporate leaders will often promote those who are similar to themselves. Many outstanding leaders have had difficulty solving this dilemma wisely; however, Jones at that time demonstrated the wisdom and vision of a true business leader. He realized that "GE is faced with three major challenges—slow increase in productivity, largely obsolete technology, and declining international competitiveness; it was having difficulty making progress" (Zhang, 2001). Based on the understanding, Jones came to the belief that the company needed someone with a style very different from his; in fact, in terms of temperament and managerial style, Jones and Jack Welch were at the two extremes. Jones was attentive to norms, but Welch often made "unusual moves." Under such circumstances, Welch became the CEO of GE in 1981. The appointment of Welch as the head of GE marked a huge change in GE's strategy for internationalizing technological innovation. Welch defined the driving forces and sources for the company's growth as globalization, Six Sigma, service, and digitalization (see Figure 2).

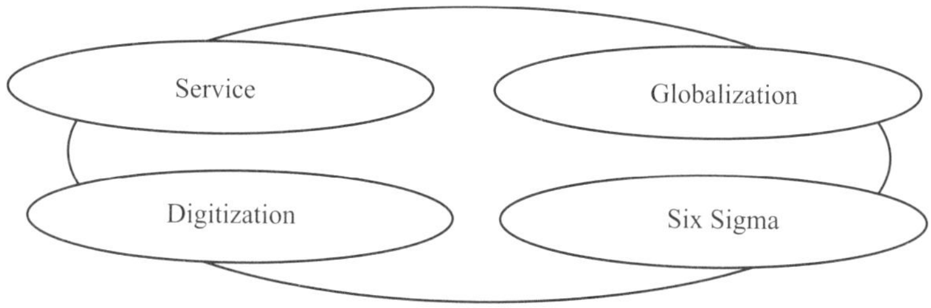

Figure 2 GE's driving forces and sources for growth as defined by Welch

After becoming the CEO of GE, Welch implemented two major changes: one is the structural adjustment and asset reallocation in the 1980s and the other is the globalization reform that started in the 1990s. At GE, true globalization was thought to be that success in the international market is only part of globalization and GE must globalize every one of its activities and, in order to go beyond market globalization, GE must make extensive use of global intellectual capital. The company must make every aspect of its business global (Luo, 2002). Under the guidance of the globalization strategy, GE's internationalization of technological innovation took on a much faster pace. In 1994, GE launched a new screw-on fluorescent lamp product, with a volume different from that of the common fluorescent lamps, power consumption reduced by two-thirds, and service life increased seven times. The development of this new product was done by a team of members from different parts of the multinational company. Overall, the design people were from the United Kingdom, manufacturing workers were from Hungary, and developers of electronic ballasts were from Ohio in the United States. The role of GE's R&D Center was limited to the formulation and coordination of the overall project.

L. S. Edelbeit, Deputy Director of GE's R&D Center, wrote in the March 1996 issue of *Chemtech*: "Our research institute can truly play its proper role only by helping the company achieve rapid development of low-cost, high-quality, and high-performance new products." (Li, 1997) To reach this goal, the company must internationalize its R&D, set up local R&D institutions, and recruit local researchers.

Under the globalization strategy, more and more business divisions moved their technology center or "quality center" from the U.S. homeland to overseas locations and started to recruit local technology talent. In China alone, GE Healthcare, GE Plastics, and GE Lighting, all have set up their "technology centers." In addition, GE also set up overseas R&D centers. For example, GE established a large-scale R&D laboratory in Bengaluru, India. It has become an R&D center next only to GE's R&D Center in Niskayuna, New York. GE's goal is that it must provide high-value products and services globally. These products and services will be designed by global talent for global market (see Figure 3).

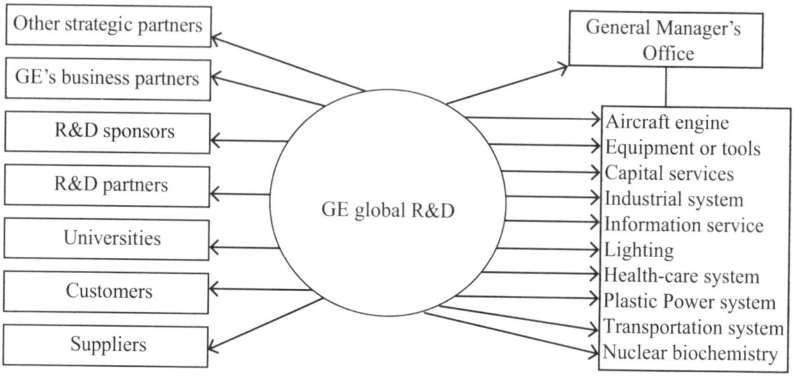

Figure 3 GE's global R&D organizations

Welch believed that only by talent globalization can the company truly get on the path toward globalization. This point has great significance to GE's business in Asia. While Asia is still an emerging market, there is good supply of high-tech talent in the region; GE has already put in efforts to find and nurture R&D resources. GE will put the biggest bet on India (see Figure 4).

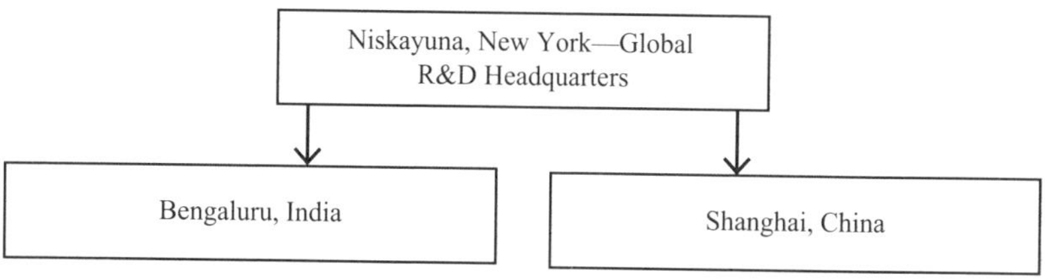

Figure 4 GE's global network of R&D institutions

At first glance, GE has already achieved the goals of its globalization strategy. However, in the face of this all-new situation, Welch proposed an ultimate goal for globalization: "global localization and local globalization." The substance of this phrase is that the company not only can localize its enterprises around the globe but also can integrate such enterprises into GE's culture and that, after gaining necessary experience, local talent do not necessarily stay in the local area and they can leave the local area—when the company's development demands so—and work in various places so as to adapt to the needs of the various developments around the globe. In high-level meetings of company R&D personnel, more and more participants are now non-U.S. citizens.

GE's internationalization strategy has also been implemented rapidly in China. In 1994, GE China was established. In June 1996, GE China Lighting Technology Center was set up in Shanghai and started to provide technology support for GE Lighting Division's joint ventures in Asia and for the suppliers of lighting products. Since then, GE has founded Plastics Application R&D Center in Shanghai, Medical System CT and X-Ray Engineering Center in Beijing, and Medical System Ultrasonic Engineering Center in Wuxi. In June 2000, GE founded GE China R&D Center in Shanghai. This marked that GE's globalization effort in China had transitioned from the phase of overseas sales of products and services and overseas procurement to a new phase of making use of intellectual capital to achieve internationalized technological innovation.

The activities of GE China R&D Center involve multiple fields. The goal of the center is to build up a world-class advanced engineering R&D institution so as to make use of China's advantages in S&T. The mission is twofold: (1) to provide support for GE's business divisions, partners, and customers, making available efficient solutions while carrying out rapid innovations; (2) to attract and develop China's S&T talent and train them into GE's future technology leaders (see Table 1 and Table 2).

Table 1 GE's R&D mission

To ensure leadership status among business partners
Products and services
To provide short-term technical support
To share technology through transactions
Source of global technology
To attract and develop GE's technology resources and leaders
To invent "game-changing" technology

Table 2 GE's two types of technologies

"Game-changing" Technology	Leading Technology
Ceramic matrix compound	Microtechnology
Solid oxide fuel cell	Photonics
High-performance polymers	Advanced propulsion technologies
Digital insurance	Light conversion technology
Advanced engine technologies	Bioprocess technology
Solid-state ultrasound	

Therefore, the center creates value for the company from three aspects: (1) to play key roles in the R&D Center's projects so as to reduce costs; (2) to provide R&D capabilities and technology support for the company's manufacturing, procurement, and service divisions; and (3) to strengthen the cooperation between the R&D center and China's various universities and research institutes (see Table 3).

Table 3 Cooperation between GE China R&D Center and universities and research institutes

Name of Institution	Project and Field of Study
SICCAS	Flash crystal and ceramics
Baotou Rare Earth Research Institute	Permanent magnetic material
Northwest Institute of Nonferrous Metals	Niobium and antimony material
Sunshine Crystal Technology Co., Ltd.	Scintillating ceramics
Zhejiang University	Power electronics and recycling
University of Science and Technology of China	Laser materials processing
Shanghai Institute of Optics and Fine Mechanics	Crystal growth
Tsinghua University	Aero-engine design and noise control

GE China R&D Center has carried out extensive cooperation with universities. Presently, it has started cooperative work with Tsinghua University, Zhejiang University, Shanghai Institute of Optics and Fine Mechanics, and so on. It has formed a complete R&D system with the R&D Center at the core and the various universities at the peripheries.

GE business divisions and their customers have greatly benefited from GE China R&D Center. First, as a provider of low-cost technology service, GE China R&D Center can lower the company's R&D costs. Second, because it is closer to the business units and customers in China, it can provide faster and more timely technical support. Meanwhile, it will also play a key role in best strengthening the exchange and communication among the business divisions in their specialized technology and research fields.

In the China R&D center of an ordinary company, the parent company usually sends over key technology managers who will perform the role of a bridge. The strategy of GE's R&D Center is not only that the technology personnel be recruited locally but also the managers be localized. Welch has consistently emphasized that the proportion of U.S. employees among the overseas staff needs to be reduced. "Foreign employees are like a stick; the stick cannot support the global opportunities." In GE China R&D Center, from top to bottom are basically all Chinese talent; this fact particularly pleases Welch (Zhang, 2001). Former head of GE China, Wang Jianmin commented on this, saying "many multinationals have made investments and set up factories in China; the products made by such factories are all for China's domestic market, or such products are developed abroad and only assembled in China. But at GE, all products are globalized. Our R&D is in China, manufacturing is in China, technology personnel are Chinese, and the production base as well as the entire chain of technological processes is all here" (Unknown, 2001).

GE's strategy for internationalizing technological innovation first attaches great importance to building up a globalized organization as a necessary condition: (i) ownership, (ii) trust, and (iii) communication of information. GE's internationalization of technological innovation has two models: (i) mentor model and (ii) technology center of excellence model (see Figure 5–Figure 7).

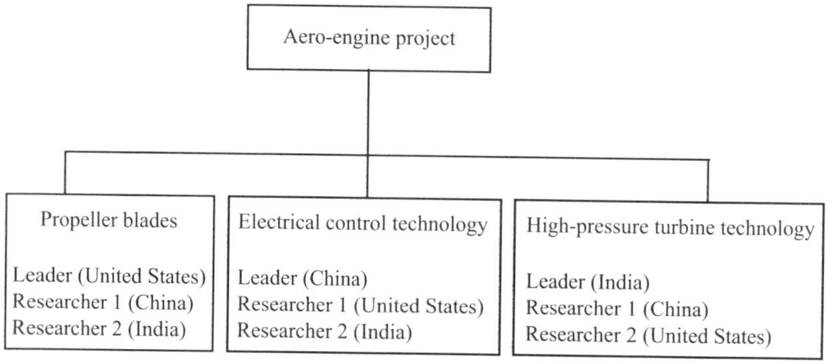

Figure 5 Mentor model for internationalizing technological innovation

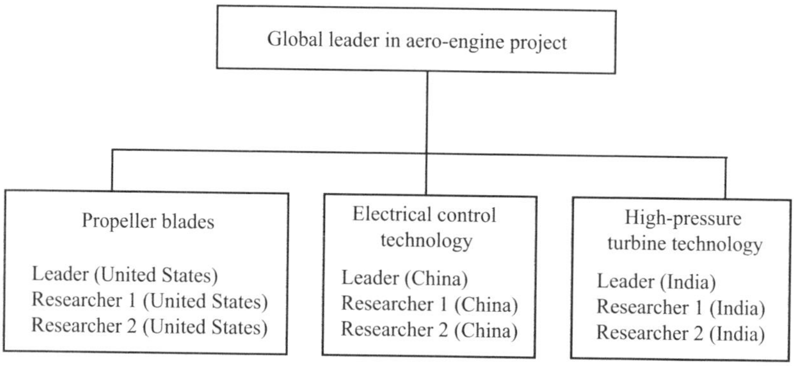

Figure 6 Technology center of excellence model for internationalizing technological innovation

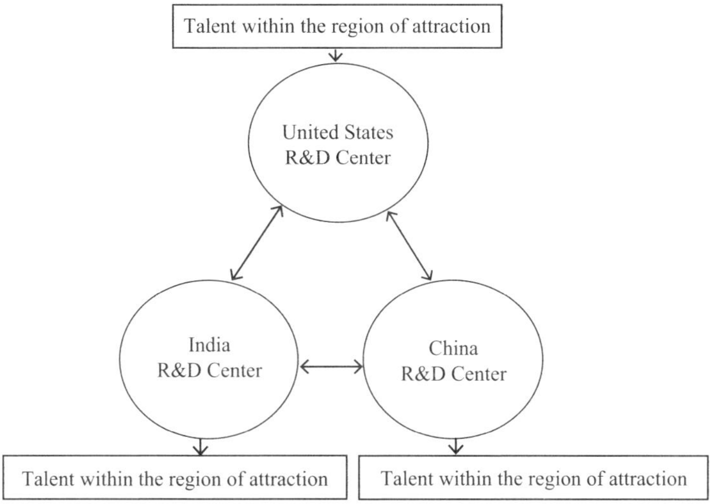

Figure 7 GE's vision for internationalization of technological innovation

3. Experience and meanings

A review of the historical path and internationalization process of GE's technological innovation reveals that the process can be divided into the following two phases.

Phase One is the one-way international flow of information and technology personnel. The characteristic of this phase is the recruiting of top-class international scientists as well as engineering and technological personnel to work at the R&D center in the United States or inviting such people to the R&D center to give lectures so as to promote the inflow of international technological information. Phase Two is when the company sets up R&D institutions worldwide, recruits local talent, and achieves the localization of R&D. It is worth noting that Welch sketched the internationalization model for GE's future technological innovation, that is, global localization and local globalization. This is a bidirectional interactive process where R&D personnel can transition between different levels within a technological innovation system. GE's innovation internationalization process reveals the following to us.

(1) GE's internationalization of technological innovation has diversified forms

It includes (i) recruiting outstanding international talent, (ii) collecting information on the frontiers of international S&T research, (iii) attracting students returning from the developed countries, (iv) participating in or hosting international conferences or international S&T forum and sharing information with partners in cooperation, (v) acquiring enterprises with high-tech talent, (vi) setting up overseas R&D institutions, (vii) working in cooperation with foreign universities, and (viii) working in cooperation with overseas high-tech companies. Among them, attracting outstanding talent has always been the approach GE attaches the greatest importance to.

(2) The model for GE's technological innovation evolved through different phases of the historical development

Phase One is to attract the talent of technologically advanced countries or talent who have already returned from those countries. One of the Phase One activities is to extensively collect international technological information. Phase Two is to set up overseas R&D institutions. Phase Two can be further divided into two stages. Stage One is after the mentor model for internationalization of technological innovation. In this case, the headquarters exerts strong business and administrative control over overseas R&D institutions. Stage Two is after the technology center of excellence model, where the various research centers around the globe are responsible for different technologies, and there are no energy-level differentiations. Then, there will be a third stage in the future of GE's internationalization of technological innovation, where there will not be a center, talent will flow freely, and "global localization and local globalization" will have been achieved.

(3) GE's model for internationalization of technological innovation is a hierarchical one

Various overseas R&D centers constitute the arteries of the

R&D network and form the core layer; universities and companies in cooperative relationships with GE constitute the ancillary layer; general universities and research institutes constitute the peripheral layer. GE's technology monitoring department will always keep an eye on this and will bring the appropriate institutions into the ancillary layer at the right time.

(4) GE's internationalization of technological innovation fully considers the local resources and the cultural and policy environment

For a research project in a location like Baotou, although the place is less developed, the research in rare earth there is in the lead; this point is neglected by many multinationals that focus on big cities.

(5) The target model for GE's internationalization of technological innovation is "global localization and local globalization"

In terms of both its breadth and the intensity of cooperation, this is a revolutionary model for internationalizing technological innovation.

(6) The main goals of GE's internationalization of technological innovation are the triad (organization, people, and information) of internationalization of technological innovation

The research institutes in different countries are different in functions. The GE R&D centers have to collect technological information from those institutions in technologically advanced countries; those in developing countries mainly provide market support. These two kinds of research institutes are positioned differently.

Case 6

Wanxiang Group's internationalization

1. Wanxiang's history of development

Wanxiang Group is an enterprise group with township enterprises forming the main body; it was founded in 1969. It is a separately planned enterprise group in Zhejiang Province and is one of the nation's 120 pilot enterprise groups and 520 key enterprises.

Wanxiang Group is functioning in the automotive parts business, which is its main business; its main products include shock absorbers, constant-velocity drive shafts, universal joint cross shafts assemblies, bearings, drive shafts, automotive rubber seals, brakes, etc. At the same time, the group fosters its business operations in trade, circulation, investment as well as investment banking, and transnational business operations. The strategic goal of Wanxiang Group is to become a large modern company with core competency and core value.

In 1969, Lu Guanqiu, through thick and thin, founded Ningwei Commune Farm Machinery Factory; like a ship, Wanxiang Group started its voyage from there and embarked on a difficult journey of entrepreneurship. Its history can be divided into the following three periods:

(1) Start-up period featuring factory workshop and diversified production (1969–1979);

(2) Growth period featuring specialization of production and modernization of management (1980–1989); and

(3) Leap-forward period featuring grouping of enterprises and internationalization of operations (1990–present).

In the 1990s, the company put forward the strategy of "big group strategizing, small unit business accounting, capital operation, and international market"; adjusted the industry structure, product structure, market structure, personnel structure, capital structure, and organizational structure; and started to seek cross-industry and cross-border development.

In 1990, with the approval of the People's Government of Zhejiang Province, Wanxiang Group was officially established. In 1994, Wanxiang created a board of directors, by which it started to transform the subordinate enterprises into companies, so that the created companies can gain independent legal status and become entities that can directly face the market competition. The board of directors is the highest decision-making body, which is responsible for major decisions and capital management. The group exercises three major functions, namely, development, control, and organization—upon the subordinate enterprises. Under this management structure, the "big when viewed from the outside with small units inside" feature took shape and the group avoided the problem of the strain-retardation-causing "dinosaur symptom" which plagued big enterprises, reduced the scope of necessary management, and strengthened its ability to develop overseas.

Relying on the advantage in universal joint, Wanxiang took "high input, sophisticated equipment, high-level talent, and high-quality products" as the guiding principles and carried out the development of series of automotive parts products; currently, it has developed product series including CVJ drive shafts, transmission shafts, bearings, car shock absorbers, and so on. In 1994, Wanxiang Qianchao stock started trading in Shenzhen Stock Exchange. In 1996, the group started asset reorganization. Wanxiang Qianchao acquired 60% of the stock in 7 companies including machinery company, special shaft company, drive shaft company, and so on; in so doing, the company used a relatively low investment and, with the fastest speed, acquired multiple stable and growing companies with annual sales over 100 million yuan and annual profit at 10 million yuan, and in a short period formed a huge production capacity and entered a period of rapid growth.

During the same period when it was developing its main business of auto parts, Wanxiang adjusted its industry structure and, using capital operations as the main approach, actively cultivated business in new industries. It has built up its supply business of automotive parts system represented by Wanxiang Qianchao Co., Ltd.; overseas sales business headed by Wanxiang's U.S. company; distribution and trade business with Shanghai Wanxiang Investment Co., Ltd. as the center; and four industrial areas, such as industry, equity, real estate, and investment banking with Wanxiang Group as the controlling body.

2. Wanxiang's management structure

There are four major companies within Wanxiang Group—Wanxiang Group Corporation, Wanxiang Qianchao Co., Ltd., Shanghai Wanxiang Investment Co., Ltd., and Wanxiang America Corporation.

Wanxiang Group Corporation is the parent company of the group, which pursues the goal of being a comprehensive trading company and is committed to developing investment banking business with capital accumulation as the objective and industrial development, equity, and real estate investment as the means. Wanxiang Group Corporation is the core company and exercises control, organization, and development functions over the subordinate units.

Wanxiang Qianchao Co., Ltd. represents the industrial companies; it specializes in the manufacturing and sales of various automotive parts and is the parts supplier of the system. Its stock "Wanxiang Qianchao" became an openly traded stock in Shenzhen Stock Exchange in 1994. In August 1997, it started supplying products to the supporting market of General Motors of the United States; and thus, after becoming China's first company exporting automotive parts to the international market in 1984, it became China's first company that has entered the supporting markets of the world's first-class OEMs. The development objective of Wanxiang Qianchao Co., Ltd. is to transform advantages in production and capital to advantages in markets and develop into a manufacturer and supplier of system parts.

Shanghai Wanxiang Investment Co., Ltd. specializes in the construction of market circulation systems. It positions its development to aim at the global market, stand on the foundation of information network, adopt a marketing strategy with service as the core, and build high-efficiency circulation systems in various business areas that are specialized, large-scale, branded, and capable of configuring various resources.

Wanxiang America Corporation is Wanxiang Group's wholly owned overseas subsidiary, headquartered in Chicago, the United States. It is responsible for the expansion of the group's transnational business and positions its development to become an international business with multiple resource allocation approaches and capabilities.

3. Wanxiang's internationalization process

Wanxiang Group has established relationships one after another with many companies in the United States, the United Kingdom, Germany, Canada, and other European and American countries by means of directly establishing networks, M&As, and equity participation and has constructed an international sales network covering more than 50 countries and regions around the globe. Established in 1994 with the approval of the Ministry of Foreign Trade and Economic Cooperation, Wanxiang America Corporation has followed the international development strategy of Wanxiang Group and has been committed to overseas expansion and to building up a Wanxiang multinational group. Wanxiang America Corporation is the bridgehead for the overseas expansion by Wanxiang Group. Wanxiang America Corporation has developed creatively overseas; one after another, it has successfully merged or acquired overseas companies, including AS of the United Kingdom, Scheller & Co., ID, LT, QA1, and UAI of the United States, by means of "equity in exchange for market, shares for market, equipment for market, market for market, and profit for market."

In 1994, the total sales of Wanxiang America Corporation were $3.5 million; by 2000, the total sales had reached $60 million, an increase of about 20 times within a short period of six years. The export business of the group has also been growing steadily year by year; in 1984, its total export was $0.2 million; in 1996, it increased to more than $50 million; and by 2000, it reached $134 million. Thus the rapid growth in export volume has greatly improved the overall profitability of the group.

M&As have helped raise the technology platform for Wanxiang Group and have enabled the company to achieve rapid integration with international standards. For example, Scheller & Co. was in possession of the world's largest pool of universal joint technology patents, and the company has been acquired by Wanxiang. As a result, the company in possession of the world's largest pool of universal joint patents is now Wanxiang.

M&As have also further promoted the internationalization of Wanxiang's personnel structure, employee flow, and employee quality. In its overseas M&As, Wanxiang usually does not gain control by means of a full takeover; instead, it usually achieves its goals with stock ownership of relative or absolute majority, which reflects the "localization" international development strategy adopted by Wanxiang Group. In this way, management team and technology professionals of the acquired company will be incorporated into Wanxiang's human resources pool, which makes its personnel structure ever more internationalized. At the same time, the exchanges between domestic and foreign enterprises, inspections, and optimizing Wanxiang's human resources as well as reallocation, all as a result of the M&As, bring about the internationalized personnel in the workflow. It can be said that Wanxiang Group has now integrated its domestic and foreign human resources management and is now able to allocate human resources globally.

After China's accession to the WTO, the key factor to the future market competition has become the competition for the core market. How to win the core market? Wanxiang's approach is to go from "going out" to "bringing in"—nurturing core technology—and from "international marketing" to "international manufacturing"—holding core markets—and eventually building up an international market system with a strong brand advantage.

Since 1984, when its universal joint products were first sold to the United States, Wanxiang has formed a developing structure of systematic development, specialized production, and serialized products as follows.

· Establish an innovation system. Since being recognized in 1996 as a national technology center, Wanxiang Group Technology Center has relied on the input mechanism, operating mechanism, and incentive mechanism as the guarantee; the R&D center, S&T information center, computer center, testing center, and S&T management center as the support; and the eight specialized product institutes as the main body and on that basis has basically established a "one-center, three-mechanism, five-center, and eight-product-line" technological innovation system. It has imported internationally advanced hardware and software systems, and its products have basically realized serialization and parameterization. Currently, Wanxiang's product development has basically met the requirement that there should be one generation of products in production, one generation in development, and one generation in reserve.

· Ensure investment in S&T. There can be output only if you have made input and large investment brings large output. Wanxiang has set up special technology development funds, which are earmarked for that specific purpose. Every year, the expenditure for R&D accounts for about 3.8% of the sales revenue of that same year. Meanwhile, it has increased the level of income distribution to R&D personnel and has implemented a "challenging salary system," that is, aside from a guaranteed salary, a research employee will draw a greater portion of his or her income from his or her project salary, which will be paid as a one-time reward when his or her project has reached its goal or will be paid continually as the project generates profits.

· Establish an external intellectual support system. First, through relevant departments of the country, foreign experts are introduced. Wanxiang has introduced a number of experts from the United States, Germany, France, the Netherlands, and so on over time to provide technological guidance and help solve difficult problems. Second, through cooperation with universities and research institutes, it has set up joint development centers. For example, it has established long-term, comprehensive, and close cooperative relationships with

Luoyang Bearing Research Institute, Tsinghua University, Beihang University, Zhejiang University, and so on. Third, it has participated in industry associations and societies so as to understand the direction of technology development and acquire industry information. In addition, Wanxiang has also established postdoctoral research stations, attracting postdoctoral fellows to work in these research stations.

For an enterprise, to go global is not just to sell its products overseas but also to give full play to its "comparative" advantages, foster its strength, avoid its weaknesses, make foreign things serve China's needs, and organically integrate its own resources with the resources of the international market. In recent years, Wanxiang again made bold explorations and tried "equity-for-market," "equipment-for-market," "profit-for-market," and other flexible approaches to acquire overseas companies. For example, it has acquired British Bearing Company and formed Wanxiang European Bearing Corporation that is responsible for its bearing business in the European region. From "international marketing" to "international manufacturing," Wanxiang has gone from passively waiting for customer orders to actively engaging in cooperation with customers, establishing a community of common interests, and ultimately attaining the goal of holding the core markets.

4. Wanxiang's internationalization of technological innovation

Wanxiang's strategic intent is, through expanding overseas, establishing R&D institutions overseas, carrying out joint development with local enterprises and research institutes, and learning from leading local users to develop the company's potential technological capabilities, improve its core technological competency, and eventually achieve the objective of stabilizing its domestic market while acquiring a part of the overseas market. We believe that, presently, Wanxiang's R&D activities are of the home country-based technological innovation type and that they are moving toward the market-driven technological innovation type; in other words, the focus of the company's technological innovation will be on market expansion and developing frontier S&T will be something in store for the future.

(1) R&D capital, scale, and tendency of R&D activities

As shown in our analysis earlier, Wanxiang's strategic intent is, through expanding overseas, establishing R&D institutions overseas, carrying out joint development with local enterprises and research institutes, and learning from leading local users to develop the company's potential technological capabilities, improve its core technological competency, and eventually achieve the objective of stabilizing its domestic market while taking a part of the overseas market. The overseas technological innovation activities are mainly aimed at market-oriented (both the domestic and overseas market) commercial development and improvement of product development as well as market-oriented reformation of technological processes. Medium-term applied research takes the next important place, and long-term basic scientific research is assigned the least important place.

(2) Main approaches and driving factors of Wanxiang's internationalization of technological innovation

Presently, the main approaches by which Wanxiang Group carries out overseas technological innovation are importing overseas technologies through obtaining technology permits, recruiting outstanding scientists and technology professionals from abroad (the appropriate scientists and technology professionals, to be precise), establishing cooperative relationships in research with overseas universities and research institutes, and trying to set up overseas R&D branches. As already mentioned, Wanxiang Group has established an R&D center in the United States (mainly supported by Wanxiang America Corporation). Being constrained by its own limited strength, Wanxiang Group acquired few overseas companies for the purpose of promoting technological innovation; so far, Wanxiang Group's several overseas acquisitions have been done mainly for expanding its overseas sales network as well as its overseas production and manufacturing network; Wanxiang Group is some distance away from the position where it can form transnational alliances for technological innovation with large companies overseas (of note, to a degree, to establish transnational strategic alliances with overseas multinationals requires that the allied parties should be of comparable strength; it also takes time to build such alliances). It is worth noting that Wanxiang Group acquired Scheller & Co., in the United States. Although to an extent the acquisition was also done for expanding its sales network, Scheller & Co. is very advanced in its technology for manufacturing universal joints and is in possession of a considerable number of U.S. patents. Thus, after acquiring Scheller & Co., Wanxiang Group has become the company in possession of the most universal joint-related patents in the United States. To the greatest extent, Wanxiang Group plans to rely on joint research with overseas universities and research institutes in its future R&D activities and at the same time plans to intensify its effort in setting up overseas technological innovation branches. Wanxiang Group plans to set up one or two R&D institutions in Europe. In addition, the group plans to intensify its effort to recruit overseas research and technical personnel, apply for overseas patents through overseas subsidiaries, establish strategic alliances with overseas enterprises, increase its overseas acquisition activities for promoting technological innovation, and lower the importance of permit purchases for importing overseas technologies (of note, in a way, Wanxiang relied on its business of OEM automotive parts manufacturing for General Motors for the early period of its growth).

We find that, the factors driving Wanxiang Group to internationalize its technological innovation are mainly responding to competition and implementing the company's development strategy, expanding and enhancing its potential technological capabilities, rapidly introducing the company's products to the local market, recruiting outstanding researchers, and absorbing advanced scientific and technological knowledge from abroad. An especially important one among Wanxiang's planned actions is about carrying out cooperative R&D projects with local enterprises and research institutes. Other factors include providing necessary technology support for local subsidiaries, making use of good local R&D infrastructure, better adapting to the special needs of consumers in different parts of the world, producing featured products, taking advantage of favorable local policies, setting up technology monitoring sites so as to learn about the competitors, determining the trend of technological development, and exploiting the development in telecommunications and information technologies. These other factors stand behind those driving factors mentioned earlier. It should be noted that Wanxiang Group believes that gaining proximity to suppliers is the least important for them among the various factors influencing internationalization of technological innovation.

An analysis of the abovementioned conditions led us to believe that this is a rational decision made by Wanxiang Group under the guidance of its comprehensive strategy. The business objectives of Wanxiang's operations are taking good care of the group's bread-and-butter business (that is, automotive parts business), thus building up its domestic market network; setting up a Wanxiang unit in China wherever there is market; stabilizing Wanxiang's leading position in the industry in China; and, on this basis, developing overseas operations and taking a share of the international market. Although Wanxiang Group has been implementing its strategy for internationalizing its business operations for ten years, there is still a long process ahead for it to achieve true internationalization. Presently, Wanxiang's internationalization has just started to gain momentum; its overseas sales revenue is still a small portion of the group's overall sales revenue, and its main market is still in China. However, Wanxiang's leaders have fully recognized that moving toward the worldwide market is the general trend. After China's accession to the WTO, on the one hand, Wanxiang will face stronger competition and, on the other hand, it will come across valuable opportunities, especially in the vast U.S. and European markets.

Although Wanxiang has already established itself as the domestic leader in the automotive parts industry, Wanxiang's technology is still not advanced enough in the worldwide market and its ability to carry out technological innovation is still inadequate. Thus, in response to the present competition in the U.S. and South American markets and in anticipation of the probably intensifying competition in China's domestic market after China's accession to the WTO, the group attaches great importance to expanding its potential technological capabilities with its R&D activities. Wanxiang has already been selling its products to the various regions of the world, but its products do not command obvious competitive advantages, and its products should be adapted to the needs in the local markets. Thus, Wanxiang attaches special importance to joint R&D projects in cooperation with local enterprises and research institutes overseas for adapting its products to the needs of the local markets and at the same time absorbing overseas S&T to meet the needs for stabilizing its dominant status in the domestic market. As it expands into various markets worldwide, Wanxiang will be faced with more intensive competition. Under such circumstances, how to ensure that the company will maintain its advantages in market competition and how to improve its own core competency have become crucial to its success. Thus, Wanxiang Group plans to further increase its cooperation with local enterprises and research institutes, recruit outstanding overseas researchers, and further develop its potential technological capabilities.

(3) Personnel selection and organizational structure

After analyzing the organizational structure and personnel selection for Wanxiang's internationalization of technological innovation, we found that Wanxiang's R&D activities are focused on building a domestic R&D headquarters to guide overseas R&D activities and that the headquarters exerts powerful control over its overseas R&D activities. This is consistent with the needs of the company's current business operations. Limited by its current status and corporate strategy, Wanxiang Group is still unable to build up a close-knit global R&D

organizational network; it is not even able to build up a close-knit R&D organizational network for certain regions of the world. However, for Wanxiang's long-term development, the company plans to build up a relatively close-knit R&D organizational network and thereby more closely link the markets overseas and absorb the knowledge of the world's S&T centers of excellence and integrate such knowledge into the company's core competency. Meanwhile, the company plans to continue the current approach of close control of its overseas R&D branches in five years, but the plan to build up such a close-knit overseas R&D organizational network within five years seems unrealistic or, at least to an extent, too hasty; it is somewhat contradictory to the plan of following the current approach of close control over its overseas R&D branches. Meanwhile, we should note that it has not been long since Wanxiang Group started its overseas R&D activities and thus the best organizational structure for such activities is yet to be found.

Localization of marketing and localization of management system are basic to Wanxiang's strategy for overseas expansion. We noted that in Wanxiang's overseas branches, local employees account for large portions of both the researchers and the managers (70%~90%), suggesting that Wanxiang has fully implemented its localization strategy in its overseas R&D activities, fully tapped the technical advantages and familiarity with the local environment of the local employees, and so far the implementation of this strategy has been quite successful. When selecting researchers for its overseas R&D branches, Wanxiang attaches great importance to the researchers' learning ability, technological competence, creativity, dedication, and skills in international cooperation; Wanxiang's approach is in line with the conclusions reached by scholars in the field of innovation research, such as Chen (2001) and Xu (2000). When selecting managers for its overseas R&D branches, Wanxiang attaches great importance to technological competency, communication skills, coordination skills, and the ability to communicate with people inside and outside the company and to obtain extensive information and to fully understand the company's strategy; the company does not give much thought about the costs involved in searching for the best managers, and the experience of R&D candidates is ranked immediately after the abovementioned important characteristics.

(4) Exchange and communication

For Wanxiang Group, the R&D branch in the United States has had great impact on the management of its R&D. Then, how did Wanxiang solve communication problems related to its R&D branches? And how did Wanxiang achieve effective communication between its branches and the company's R&D headquarters? Our survey showed that Wanxiang attached great importance to the use of information technology for exchange and communication and that according to Wanxiang, even though information technology cannot provide the effectiveness of face-to-face communication, it can still help achieve satisfactory results. Wanxiang's network across the world has also been providing feedback about market dynamics and been partially performing the function of technology monitors, and the sales branches on the network have been exchanging information and communicating with the company's headquarters and R&D headquarters via information technology. In implementing R&D internationalization, Wanxiang attaches great importance to the exchange and communication between the supervisors of R&D branches in the United States and the company's headquarters; the goal is to ensure that the R&D branches adequately understand the company's strategies and intentions. In addition, Wanxiang values regular contact between managers and R&D personnel of overseas R&D branches and encourages the branches to establish a culture of open information exchange. Meanwhile, overseas R&D branches also organize various formal and informal seminars, carry out research personnel and management staff job rotation, and offer various kinds of language training where participants introduce different cultures, social backgrounds, and customs of their countries, so as to enhance the exchange and communication among the employees of different nationalities. It is worth noting that according to Wanxiang the most important factor affecting the effectiveness of exchange and communication is the level of trust among R&D personnel. In the practice of its R&D internationalization, Wanxiang has not been very satisfied with the effectiveness of the exchange and communication in the R&D activities and expects to make improvements in this area.

(5) Likely impact of China's accession to the WTO on Wanxiang

After joining the WTO, the basic trend for the development of China's economic system, as well as trade and investment liberalization, is further opening up, and thus, China's economy and industries are bound to be impacted and affected. Considering the question of how much China's economy will be impacted after China's accession to the WTO, different scholars hold different opinions. Some scholars believe that the negative impact of China's accession to the WTO will be greater than the positive impact, but there are also scholars who believe that we should not exaggerate the negative impact of China's accession to the WTO and that the actual situation may be exactly opposite—the positive impact will be greater than the negative impact. In our opinion, China's accession to the WTO will impact different industries and companies differently. However, in this book, we will not discuss at the macro level whether it will do more good or more harm; we will only try to discuss what impact China's entry into the WTO will have on the technological innovation of Chinese enterprises.

Of note, Wanxiang Group believes that after China's accession to the WTO, changes in intellectual property policy and trade policy will have a greater impact on Chinese enterprises; however, changes in related policies concerning taxation, finance, fiscal budget, S&T, and innovation will not have too much impact on the enterprises. The greatest impact on enterprise technological innovation will be that it may block most of the imitation innovation and, as a result, enterprises will have to increase investment in R&D and acquire their own intellectual properties and thus enterprises will be facing tremendous pressure of R&D funding. Meanwhile, to an extent, it will prohibit enterprises from independently developing software products with the same functionalities and carrying out reverse engineering, increase the difficulty for Chinese enterprises in their effort to acquire technologies from foreign enterprises, and augment the production costs of China's technology industry. Wanxiang does not believe that, after China's accession to the WTO, enforcement of provisions related to intellectual property rights will further increase the monopoly of developed countries in knowledge supply or that developed countries will necessarily slow down their technology transfer to Chinese enterprises. Wanxiang believes that, after China's accession to the WTO, China will implement the Subsidies and Countervailing Measures Agreement[1]; afterward, China will have to make related adjustments to its many S&T and finance policies and will have to cancel various government grants, loan discounts, VAT, export tax rebates, subsidies, and financial incentives. The impact will be minimal to Wanxiang Group because since its establishment, Wanxiang has nearly never benefited from any preferential policies. After its accession to the WTO, China will enforce the related TRIMs agreement and cancel the local content requirements, balanced trade requirements, domestic sales requirements, technology transfer requirements, foreign exchange restrictions, and so on; these changes will not have much impact on Wanxiang either.

[1] This agreement prohibits various subsidies conditioned upon export performance and import substitution and, at the same time, strictly limits various subsidies for R&D, with an exception for basic research, industrial research, and precompetitive development activities for subsidies. This agreement also stipulates that subsidies for industrial research and precompetitive development shall not exceed 75% and 50% of its reasonable cost, respectively.

Case 7

ZTE's internationalization of technological innovation

Internationalization of technological innovation is for an enterprise to extend its innovative activities abroad—by directly establishing overseas R&D institutions, through M&As, by forging international technology alliances, or by other means. It is also a new paradigm of technological innovation characterized by globalized resource acquisition, multinational innovative talent and network-based innovative organizations. It enables the formation of an enormous knowledge transaction network connecting the internal system of the enterprise with the various external organizations and changes the situation when the enterprise can only obtain necessary technological knowledge internally. An enterprise can draw from a broader range of external knowledge and data to build up its own technology and knowledge structure.

After World War II, with the rapidly developing trend of

economic globalization and ever-intensifying international competition, big multinationals in advanced countries—to meet the demands arising from the complexity of the markets worldwide, the wide variety of products, and the different preferences of consumers of different countries; to make use of the special technological resources of the various countries worldwide; and to contain the costs and risks involved in the process of new product development—walked away from the singular approach of home-country-based technological innovation. In accordance with the comparative advantages in talent and technological strength and in technological infrastructure of the host countries (Lin & Chai, 1998), they started to carry out globalized technological innovation activities one after another. Internationalized technological innovation has already become the dominant design and winning weapon for today's multinationals to carry out technological innovation and to engage in international competition.

Faced with the trend of economic globalization, especially the new trends of global competition after China's accession to the WTO, Chinese enterprises urgently need to find a new path to rapidly improve enterprise technological capabilities and effectively respond to international competition. We can learn some useful lessons from ZTE's experience in its process of internationalization of technological innovation.

ZTE is one of China's largest listed telecommunications equipment manufacturing company; it has a full range of communications products in areas such as mobile, data, optical communications, exchange, access, and video communications; and it is one of the major equipment suppliers for the construction of China's telecommunications infrastructure. "Technology innovation is the eternal theme of enterprise development, and the ability to engage in continued technological innovation is the source of sustainable enterprise development"—this is ZTE's mission statement.

Especially since 1995 when ZTE began to implement the three strategic shifts, internationalization has ever become the most important part of the company's development strategy. At the same time, when implementing market internationalization, ZTE was actively pursuing internationalization of technological innovation, broadening its access to technology, and expanding its technological innovative sources. It was also paying close attention to promoting international technological cooperation and actively establishing technology alliances, and it has made full use of the technology and talent advantages in developed countries and the spillover of the technological innovation and has explored a path for developing technological advantages with global innovative resources.

1. Internationalization of innovative sources

Technological innovation starts from idea formation. The ability to acquire innovative ideas and to quickly and economically identify innovation opportunities is the prerequisite for the success of a company's technological innovation. In addition, as with any activities of S&T, technological innovation is a flow of endless knowledge generation and continued development. This "knowledge flow" is also an "information flow." Only when useful information can be acquired in a timely manner will technological innovation activities be effectively carried out (Chen, 2001).

ZTE's internationalized technological innovation first needs to address the issues related to the acquisition of innovative ideas, innovative information, and technology. Currently, it has basically established a set of methods for absorbing and obtaining foreign technologies (see Figure 1).

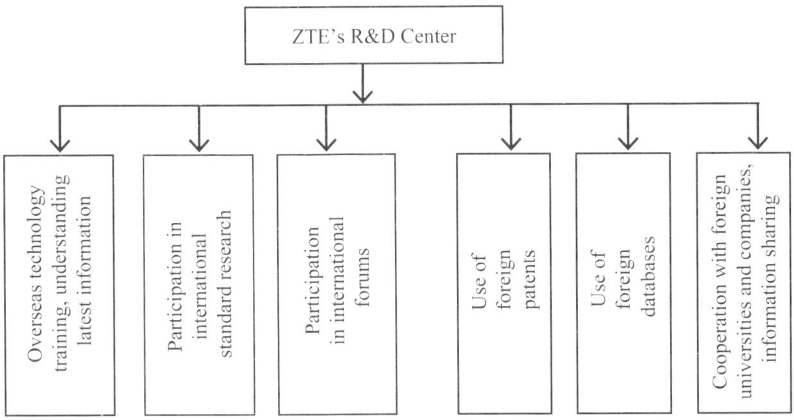

Figure 1 Main approaches of ZTE to obtain foreign technology

(1) Active participation in international standards research activities

The industry standards organization is responsible for developing the technical specifications in the industry; it provides ideas for enterprise technological innovation, guides the direction, and determines the outcome of such innovation.

ZTE regards the participation in the activities of international standards organizations as an important approach for obtaining information related to technological innovation. ZTE regularly participates in the activities of various international standards organizations and conferences such as ITU, 3GPP, 3GPP2, CDG, NPF, DSL (see Table 1).

Table 1 Examples of international and domestic standards organizations or forums that ZTE has joined

As an independent member			Participating in activities
3GPP2	CDG	NPF	ITU
AIC	DSL	CTSI	3GPP

Through participation in the activities of international standards organizations, ZTE has been informed of the latest trends of the industry technology standards, broadened the idea about its technological innovation, been assured of the direction of innovation, and gained the initiative in technology development. As Kan Yulun, Deputy Director of ZTE Technology Center, has said, "Active participation in the activities of international standards organizations, for ZTE Corporation, is a process of coming in and going out." It is reflected in the following two aspects: (1) to participate in the activities of the international standards organization, actively learn and understand the trend of technological development in the technical field, and understand the needs of the market and customers, so as to avoid detours and shorten the time for R&D; (2) to introduce the advantages of the company's product development process to the international standards organization and thereby to establish its technological advantages and leadership status in the technical field.

(2) International forum

ZTE attaches great importance to participating in international academic conferences and forums in its technical field and uses such activities as the means to promote technology exchange and transfer between the company and other internationally advanced enterprises. On October 29, 2001, ZTE hosted the TSG-S and SC/OP conference organized by the third-generation mobile communication international standards organization 3GPP2. Through this conference, ZTE gained further understanding about the intellectual property policies of the Telecommunications Industry Association of the United States, R.O. Korea's Telecommunications Technology Association, and Japan's Communications Industry Council.

(3) Foreign patents

ZTE's first step in technological innovation is to vigorously introduce technology from the outside and then to imitate such technology and innovation. Buying patents from overseas patent agents in the United States, R.O. Korea, and so on has become an important way for ZTE to acquire technology. This greatly enhanced the pace of ZTE's development work in the field of CDMA technology.

(4) Overseas databases

ZTE recognizes that databases are an excellent source of inexpensive technology. Using a database, an enterprise can obtain a large quantity of S&T information within a limited time and enrich its stock of innovative ideas. ZTE has established contact with technology networks operated by companies in the United States, R.O. Korea, and Finland and with other international databases and has created conditions for their researchers to use such technology databases.

(5) Overseas technology training

ZTE has established a three-level training system. To meet its various needs, the company often sends employees to foreign companies for field training and thereby provides them with opportunities to have direct contact with technologies, operations, and businesses of foreign companies, and in so doing, the company has improved the skills of its employees, broadened its horizons, and directly enhanced its ability to carry out technological innovation.

2. Establishing technology alliances and achieving complementary advantages

Technology alliances with universities, research institutes, and other companies can help both parties to achieve complementarity of innovative resources and enable them to carry out rapid technological innovation. Especially for companies engaged in global competition, technology alliances can help them obtain resources that they do not have, respond to rapid changes in technology worldwide, and ultimately win in global competition through rapid technological innovation; thus, the establishment of such alliances has become more important. Engaging in extensive cooperation with foreign universities, research institutions, and companies and carrying out a production–learning–research integration strategy are the main approaches adopted by ZTE to achieve internationalization of technological innovation.

ZTE has not only worked with major universities in the Chinese mainland, but it has also signed "ZTE Telecommunications Research Fund," an agreement of cooperation in major research projects, to establish business incubators. In the operation of the incubators, ZTE follows the "mutual respect" corporate culture and has worked with other parties to achieve complementary advantages and mutual trust, and it has also worked to merge the cultures of the two sides in cooperation, so as to continue with the incubation work of the project in a friendly fashion.

ZTE has also been in extensive cooperation with well-known overseas enterprises, building technology alliances so as to achieve breakthroughs in technological innovation capabilities. In August 1998, ZTE and TI of the United States, entered into a cooperation agreement; they were to set up TI-ZTE DSP joint laboratory in Shenzhen, which would be responsible for tracking, introduction, and development of DSP technology and for DSP training and promotion among the company's developers. In August 1999, ZTE and Motorola worked together and established a joint laboratory in Nanjing and then started extensive cooperation in data communications, mobile communications, and so on to jointly explore new telecommunications technologies.

In May 2002, ZTE and Agere Systems signed an agreement to set up "Agere-ZTE Telecommunications Technology Joint R&D Center". Both sides appointed their representatives to form the management team and worked together to provide personnel training and technology support as well as necessary R&D equipment and facilities, and Agere provided new products and related evaluation equipment. The joint R&D center focused on developing applications for the products provided by Agere Systems and on core technologies suitable for both the Chinese market and the international market.

The correct selection of alliance partners is the first issue to be considered by ZTE in its external cooperation. ZTE's focus is on selecting alliance partners with a strategic vision so as to carry out long-term cooperation; in this way, the two sides will have more opportunities to engage in fair exchange of information and more extensive technology transfer. The cooperation with Agere Systems, to a great extent, reflects the characteristics of the technology alliances entered into by ZTE. Agere Systems is the world's well-known supplier of telecommunications semiconductor and is on a leading position in the fields of optoelectronic devices and integrated circuits. In order to achieve breakthroughs in optoelectronics, integrated circuits, and data transmission technologies, ZTE actively sought cooperation with Agere Systems. In order to accomplish its strategy to enter the Chinese market, Agere Systems was also in desperate need of finding a Chinese enterprise to carry out cooperative research. Such strategy-based needs rapidly brought the two parties together.

Targeted and close cooperation is a characteristic of ZTE's cooperation effort, and mutual trust and effective communication form the foundation for the successful operation of the technology alliances. People at ZTE believe that trust is the basis of cooperation and that effective communication among cooperative members facilitates the attainment of the goals of cooperation for both sides. By working together, ZTE acquired technology and enhanced its core competencies and Agere Systems found a springboard for leaping into the Chinese market; thus, they achieved a win-win situation.

3. Going abroad and setting up overseas R&D institutions

Establishment of overseas R&D institutions is the most direct manifestation of internationalization of technological innovation. Generally, by setting up overseas R&D institutions, a company can gain the following advantages: the first advantage is the ability to meet the special needs of the various customers worldwide, to adapt to local production conditions, and to achieve technology localization in order to support the manufacturing enterprises of the company in the host country and effectively configure enterprise capabilities; the second advantage is the ability to use the institutions as a window of information and through this window to track and obtain technologies from the host country and from business competitors and to benefit from the spillover effect of local R&D and use what is gained as the supplementary source of specialized technologies; and the third advantage is the ability to get close to knowledge centers, to make use of technology talent and research environment of the host country, to lower R&D costs, to promote the generation and diffusion of innovative activities, and to raise the overall technological capabilities.

The establishment of R&D institutions marked a substantial step in internationalization of technological innovation by ZTE.

In June 1997, ZTE set up a research institute in Silicon Valley to make use of its advanced technology and first-class talent and to track the world's latest technology in the information field. The successful development, commercialization, and large-scale export of ZTE GSM mobile communication system, all accomplished within a short period of time, are attributed to the strong support from this research institute.

In September 1999, ZTE Korean Holding Company was established, which was mainly engaged in the R&D of CDMA mobile phones and in organizing the production in R.O. Korea and China. Within less than a year from its establishment, the institution successfully developed the world's first CDMA mobile phone with separate phone and card.

ZTE has set up R&D institutions in R.O. Korea and the United States (New Jersey, San Diego, and Silicon Valley), where technology and talent are concentrated; there are also R&D institutions established in China apart from the abovementioned institutions, thus forming the initial layout for its global R&D network.

Major objectives of ZTE for setting up overseas R&D institutions are to make use of the S&T talent and the research environment of the host country and to minimize the costs of R&D. For human resources, overseas R&D institutions directly recruit local technological and management professionals so as to achieve localization of talent in the host country. Meanwhile, effective motivational and communication measures should be taken so as to effectuate synergy among the employees of different cultural backgrounds.

Establishing information window—tracking as well as acquiring technology from the host countries and competitors—and getting close to knowledge center as well as gaining technology spillovers from the local R&D branches and using overseas branches as supplementary sources of specialized technology are the two additional objectives of ZTE for establishing overseas R&D institutions. Therefore, ZTE attaches great importance to establishing overseas R&D institutions in places where there are abundant talent of cutting-edge technology, complete technological infrastructure, high degree of business environment protection, and active technological innovation, as well as in the technology centers of excellence overseas.

The main tasks of overseas R&D institutions include two parts: (i) to recruit the world's best talent and make use of the local research infrastructure so as to build up the company's innovative resources; (ii) to track the world's latest information technology and gain access to the source of global technological innovation; and (iii) to engage in frontier research such as soft switches, CDMA1X upper-tier protocols and so on so as to provide the company with long-term technology support.

Overseas R&D institutions carry out research projects according to the plan laid out by the company's headquarters, and they are vertically linked to the headquarters. At the same time, they engage in a few cooperative projects with each other on the basis of arrangements made by the company's headquarters; thereby, the entire R&D system constitutes a star structure with the headquarters as the center and the four overseas R&D institutions as branches. In addition, ZTE's goals are to construct a globally integrated R&D network and ultimately achieve the globalization of its innovative sources and resources, internationalization of its innovation organizations, and integration of its innovative process and network; with such a system in place, the globally integrated R&D network will facilitate internationalized production and operations that will drive globalized R&D.

4. Hybrid management method combining top-down method and bottom-up method

Geographically scattered and network-based global R&D technology alliance has changed the company's organizational structure and processes and made management issues more complicated. Generally, for decision-making related to overseas R&D institution, there are three modes: top-down planning method (absolutely centralized), where all decisions are made by the headquarters and the overseas R&D units are only responsible for implementation; bottom-up planning method (totally liberalized), where decisions are mainly made by the overseas R&D units themselves and the company's headquarters is only responsible for supervision; and hybrid planning method, which is a combination of the top-down method and bottom-up method.

ZTE adopts the hybrid planning method for managing its overseas R&D units (see Figure 2 and Table 2).

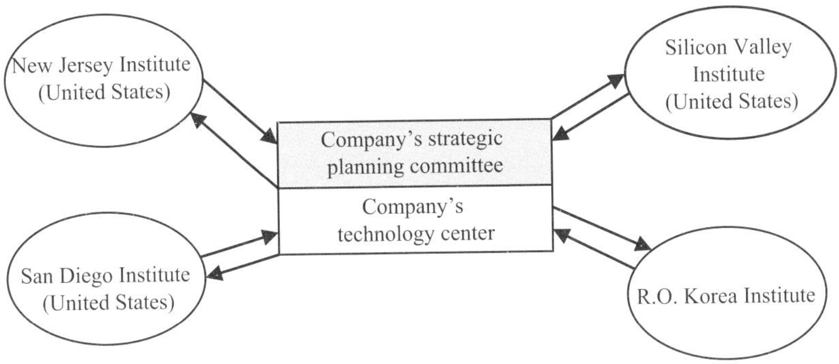

Figure 2 Schematic diagram of ZTE's overseas R&D management

Table 2 Comparison of ZTE's approaches in managing overseas R&D

Major tasks in R&D management	ZTE (Hybrid)	Top-down	Bottom-up
Setting goals for overseas R&D	Company's technology center	Higher level unit	Overseas manager
Planning of overseas R&D projects	Company's technology center and overseas manager	Higher level unit	Overseas manager
Monitoring and evaluation	Company's technology center	Higher level unit	Supporters from multiple sides
Sources of overseas R&D funds	Company's headquarters	Higher level unit	Supporters from multiple sides
Types of channels for R&D exchange and communications	Formal and informal	Hierarchical, formal channel	Equal, informal channel
Nationality of overseas R&D manager	Home and host countries	Company's home country	Host country

The highest body within the company for decision-making and implementation is the company's strategic planning committee and its technology center. The company's strategic planning committee develops plans—on the basis of industry development and technology trend—for major decisions, such as new areas for the company's medium- to long-term technology development, product R&D in the new areas, company's core technology and basic research, and entry into or withdrawal from certain industry or certain line of product; compiles technology development planning advisory report for the next three years; and provides guidance for product planning and R&D on the basis of the advisory report. The advisory report will be updated every quarter by rolling forward and followed up by the company's technology center. The four research institutes located in Silicon Valley, San Diego, and New Jersey of the United States, and Seoul of R.O. Korea belong to the technology center.

The strategic planning committee and technology center of the company plan the functions of the overseas R&D units; the main tasks of the overseas R&D units are to track the latest technological achievements of the world and the direction of the development of cutting-edge technologies and to accomplish works such as technology tracking, leaping over, preliminary research, and so on. According to the plan of the company's strategic planning committee, for strategic projects affecting the company as a whole, decisions are mainly made by the company's technology center, the overseas R&D units are responsible for the implementation, and the management at the company's headquarters monitors and evaluates the decisions and implementation activities. For other projects, decisions will be made mainly by the overseas R&D units themselves, and the company's technology center is only responsible for granting approvals and evaluation.

The selection of leaders for overseas R&D units is the most important issue for ZTE in managing its overseas R&D units. The requirements of leaders by the company for its overseas R&D units include the ability to handle cross-function and cross-discipline information and to control information flow, the ability to correctly make use of the experience within the company about running the entire product development process and to maintain project continuity, the ability to bridge cultural differences and to reconcile the needs of different people, the ability to engage in effective communication and coordination and to maintain creativity and high efficiency of the organization.

The company grants considerable autonomy to the overseas R&D units in project planning, human resource management, financial resources, and so on. Managers of the overseas R&D units have the power to make decisions on the planning of ordinary projects, recruitment, and use of funds. Management of the company's headquarters supervises the operations of the various R&D units through annual performance appraisal, so as to lower the costs and improve efficiency.

When the top-down method is adopted, managers in higher level units command greater power and bear greater responsibility for R&D management around the globe. This method can ensure the consistency of overseas R&D projects and the company's strategy and also ensure the effective use of R&D resources. However, due to the strict control over them, overseas R&D laboratories will be lacking in initiatives and creativities; the bottom-up method shows the opposite characteristics. Overseas R&D managers and researchers work together to set the goals for their own overseas R&D projects. Higher level managers merely provide minimal guidance and place minimal restrictions. This method will lead overseas laboratories to have greater flexibility and initiatives, but the goals of overseas R&D projects may not be consistent with those of the company's strategy.

ZTE's hybrid method, which combines the top-down method and the bottom-up method, has advantages of both methods. It not only ensures that the goals of overseas R&D projects are consistent with those of the company's strategy but also ensures that overseas laboratories have greater flexibility and initiatives; thus, this hybrid method is more conducive to the success of R&D projects.

5. Road of internationalization of technological innovation for Chinese enterprises as seen in ZTE's experience

In today's world, internationalization of technological innovation has become the new paradigm for enterprise technological innovation. The development process of internationalization of technological innovation shows that in the process of carrying out strategies for global competition and global innovation, multinationals in general have gone through a two-step process—the first step is market internationalization and the second step is R&D internationalization.

Generally, Chinese enterprises are presently operating at a low degree of market internationalization, and at the same time, they face tremendous competitive pressure from large international companies with great technological power; thus, Chinese enterprises cannot follow the well-traveled path by multinationals to achieve market internationalization first before carrying out internationalization of technological innovation. ZTE's experience in its process of internationalization of technological innovation provides us with some good lessons as follows.

(1) Driving market internationalization through internationalization of technological innovation

In global competition, according to the actual situation that, when compared with large international companies, ZTE started late and was at a low degree of internationalization with regard to its business operations; however, it walked a new path of driving market internationalization through internationalization of technological innovation. Through globalized innovative sources, multinational innovative talent, and network-based technological innovation organization, ZTE has on the one hand improved its own technological innovation capabilities and on the other hand become informed of the technology trend and market trend in the world's telecommunications industry. The improvements in technological capabilities have become the foundation for the company to expand into the international market. The successful development, commercialization, and large-scale export of ZTE GSM mobile communication system, all accomplished within a short period of time, are attributed to the strong support from its research institute based in Silicon Valley, the United States.

(2) Choosing the overseas technology talent, monitoring and obtaining the technologies of the host countries and competitors as the main goals

The objectives pursued by multinationals when carrying out internationalization of technological innovation can be categorized into three types: (i) to meet the special needs of the various customers worldwide, adapt to the local production conditions, achieve technology localization, support the manufacturing enterprises of ZTE in the host country, and effectively allocate ZTE's capabilities; (ii) to set up a window of information, monitor and obtain technologies of the host countries and competitors, and benefit from technology spillovers of the local R&D as well as use such spillovers as a different supplementary source of specialized technology; and (iii) to gain proximity to knowledge centers, make use of the technology talent and research environment of the host country, lower R&D costs, promote the creation and diffusion of innovative activities, and improve the company's overall technological capabilities. Based on the assessment that its weakest area in international competition was in its technological capabilities, ZTE chose the strategy for internationalizing technological innovation by making use of overseas technology talent as well as tracking and obtaining technologies of the host countries and the competitors as its main goals.

(3) Overseas R&D institutions are mainly located in the world's knowledge centers and technology centers of excellence

ZTE's internationalization of technological innovation is mainly of the technology tracking (or acquisition) type and resource seeking type. Mainly because of the need to tap into the supply of technology talent in the host countries and take advantage of the technology facilities and research as well as the business environment and the need to engage in cooperation with local universities and research institutes overseas, ZTE's overseas R&D institutions are generally set close to prestigious universities or research bases of the same type.

ZTE's experience will guide Chinese enterprises in the right direction in their efforts to carry out internationalization of technological innovation and to effectively engage in international competition.

Case 8 3M's global innovation

1. About 3M

Minnesota Mining and Manufacturing, referred to as 3M, was founded in 1902. The company is located in the rural areas of Minnesota, the United States, and is the world's most innovative company. The century-old 3M is one of the 30 stock companies that make up the Dow Jones Industrial Average; it is also a component of the Standard & Poor's 500 Index.

3M has its headquarters in Saint Paul and is a company with a number of areas of technological expertise; it is a leader in different markets, namely, health care, safety, electronics, telecommunications, and so on. 3M continuously develops new products on its over 30 technology platforms (see Table 1); the company has about 60,000 kinds of products and more than 40 business units in the following six market areas: transportation, mapping, and security; health care; industrial products; consumer and office products; electronics and communications; and specialized materials.

Table 1 Technology platforms of 3M

Adhesive	Pharmaceutical import technology	Medical equipment	Electronic equipment and software
Paint	High cleanliness paint	Foundry equipment	Fluoride and fluoride-containing polymers
Dental consumables	Specialty compounds and polymers	Optical control equipment	Ink and pigment
Luminous electronic component	Ceramic products	Polymer production technology	Micro replication technology
Film	Synthetic material	Radioactive processing technology	Nonwoven products
Medical information	Diamond film	Surface renovation technology	Micro-particle comminution technology
Mechanical locking system	Motor system	Chemical raw materials	Filtration and separation materials
Mine crushing equipment	Filtration and insulation materials	Copper interface	Skin care products
Fiber optic cable and connectors	Image technology	Monitor	Vibration control technology

Source: Gundling, E. *Fertile Soil for Innovation—3M's Innovation Mechanism*. Chen, X. S., Chi, J. C. & Zhang, H., trans. Beijing: Huaxia Press, 2001.

In the past few years, 3M was affected by factors such as economic recessions in Asia and Latin America and its profit declined. However, 3M was still fully confident and believed that such economic difficulty was just a little "atmospheric disturbance," and not a "sea storm." The management team strengthened the management of internal operations and efficiency of the company, but they did not resort to extreme measures to improve the company's short-term benefits. The company is still in an atmosphere of freedom, patience, and personal initiative. Innovation is always a worthy cause and 3M is still an enterprise that has the most active innovative programs and is the most capable in adapting to the market competition of the future.

2. Global innovation—3M's advantages

At a higher level of summarization, there are two main ways to engage in competition: one way is to optimize production resources so as to gain possible profit margin in the market; the other way is through innovation so that the company can introduce innovative products to the market with absolute advantages over those from other companies within a short time and thereby disrupt the existing market balance. Schumpeter refers to "competition through innovation" as explosives that destroy an entire building and "competition through efficiency" as merely "knocking at the front door." 3M is undoubtedly an expert in opening up new markets.

The goal of 3M is to be the world's most innovative company in the markets it serves. Therefore, "to other enterprises, innovation is an important part of the strategy; but to 3M, innovation is the strategy." (William Coyne, Senior Vice President of R&D, 3M Corporation) Along with its goal of being "the most innovative company," there are efforts by each part of the company, high and low, domestic and abroad, to integrate innovation into the mainstream of the company.

(1) Mutually supporting domestic and international innovation

In 3M, there are three different types of innovation (see Table 2). 3M's research institutes in the United States and its overseas institutions play different roles in different types of research.

Table 2 Three different types of innovation

Innovation type	Function/purpose	Customer relationship	Work direction	Features
Type A	Driving a new business or industry	Satisfying a not-yet-known need: new customers and new markets	Research	First the solution and then the problem
Type B	Shifting the foundation of the competition and providing a competitive advantage	Existing customers and new uses	Development	First the problem and then the solution
Type C	Extending product life and increasing fixed asset value	Existing customers and existing uses	Market	

Source: Gundling, E. *Fertile Soil for Innovation—3M's Innovation Mechanism*. Chen, X. S., Chi, J. C. & Zhang, H., trans. Beijing: Huaxia Press, 2001. Reorganized based on relevant parts of the book.

Type A refers to the new technology development for ten or more years in the future, and the innovation objective is to form a new industry. Type B refers to the advanced technology development, and the main task is to adapt products to the new uses by the existing users and to provide the company with competitive advantages in the existing field for the next three to ten years. For Type C, the main task is to conduct product development, product control, and technological services so as to extend the market life of the existing products.

The central research laboratory of 3M in the United States is mainly committed to long-term technology development so as to accomplish fundamental innovation; technology centers of 3M (the original advanced laboratories) are responsible for developing various technological platforms, while looking for new uses for existing products; branch laboratories of 3M are closely linked to customers and existing products and work to meet the current needs. Among 3M's overseas subsidiaries, 32 have laboratories that work in support of the company's business plan; the functions of these laboratories involve technology support, manufacturing support, product improvement, product development, and some technology development. All the other international subsidiaries also provide customers with technology support with the help of either local service engineers supported by different subsidiaries or from regional or U.S. laboratories. For units and subsidiaries geographically located at the market forefront, in comparison with the research institute in Saint Paul, the work focus is not to invent new technologies but to make full use of the company's existing resources to meet the needs of the local customers. Over the past decades, the standard model has always been hatching new products in Saint Paul, then to put them to the U.S. market, and eventually to introduce them gradually to the customers of the company's subsidiaries (see Figure 1).

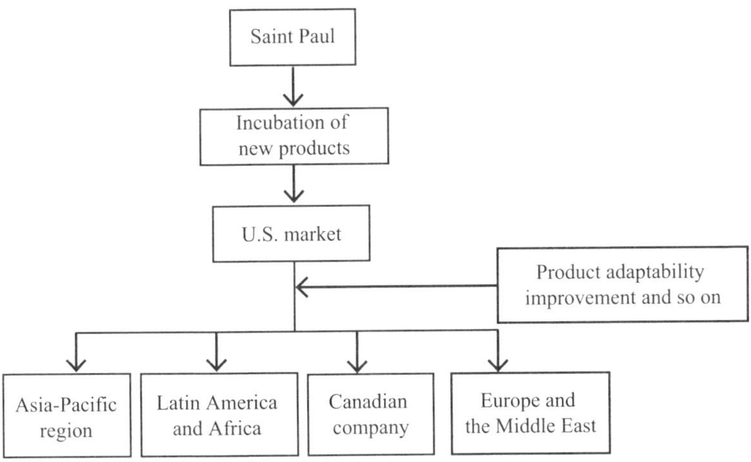

Figure 1 Standard model for 3M's product development

In recent years, a new and different model is taking shape. In a wide variety of innovative areas, all the main branch companies of 3M are moving toward the direction of closer and more intensive cooperation; the various branch companies around the globe are now cooperating in the design and development of new products, and they launch new products in the key markets around the globe at the same time (see Figure 2).

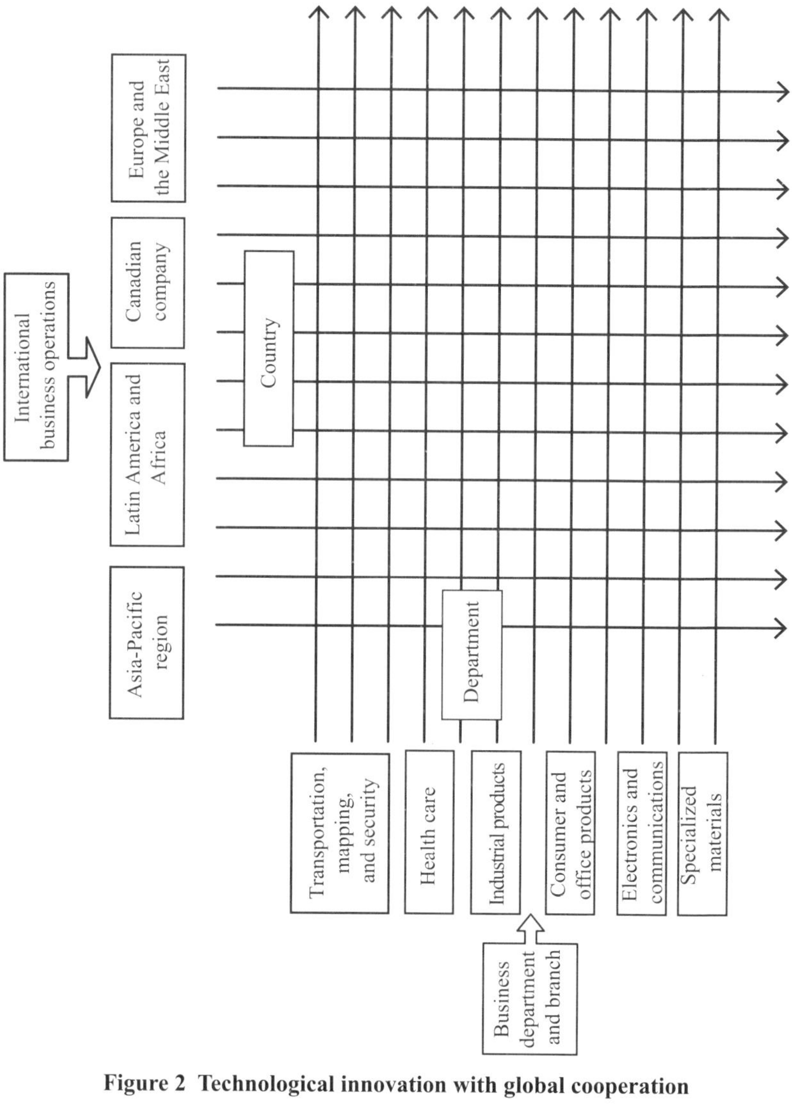

Figure 2 Technological innovation with global cooperation

(2) Development process of 3M's overseas subsidiaries

In 1929, 3M started selling products outside the United States. In 1951, it set up the first international operation and started to invest directly. Now, 3M's products are sold in nearly 200 countries, and 3M has established subsidiaries in more than 60 countries.

When 3M begins to develop business in a new location, it usually starts from selling through agents or distributors (see Figure 3). Once sales potential in the location has been established, the company will develop a local unit that directly recruits employees and sets up its own sales teams. Soon after that, 3M will be delivering large quantities of materials and will process them into localized products; then such processing will turn into all-around large-scale production, which will churn out more customer-centered products; such a strategic arrangement greatly promotes innovative activities in various branches. When a production base reaches a certain scale, it would take on the core function of engineering technology and product development and the local employees will be empowered to develop more new products and thereby to extend the various needs of the customers. Localized production is the key to gain market share. Of all the products sold in non-U.S. markets, 70% are made outside the United States; more than 50% of all 3M sales revenue comes from sales by the overseas subsidiaries (53% for 2001 and 54% for 2002), and the company is still investing heavily in this regard.

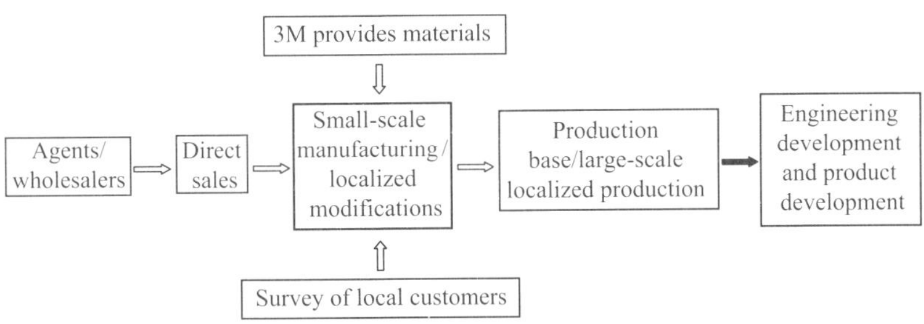

Figure 3 Development process of 3M's overseas subsidiaries

A rich and wide range of products and a globalized subsidiary network have helped to create a global product marketing model for 3M's development known as the "cascading effect" (see Figure 4). Among the various subsidiaries, as compared to the direct input from the U.S. subsidiaries, the experience of a nearby subsidiary in a leading country is usually more helpful for business development in a particular country. For a relatively new branch company, it is not necessary to reinvent the wheel. This provides a free space for employees and subsidiaries that are far away from Saint Paul. In addition, local innovations can also cross national boundaries and be channeled to the markets with similar needs in other countries, thereby gaining even greater return on investment for the company.

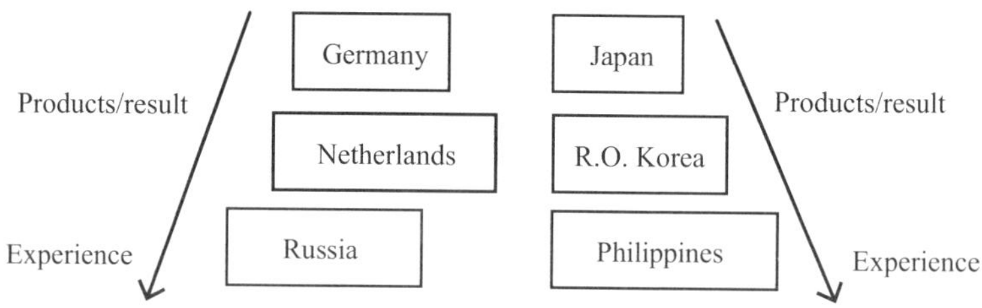

Figure 4 Cascading effect

(3) Fields of 3M's internationalized innovative activities

Although technological innovation by 3M's international subsidiaries accounts for only a small portion of that by the entire 3M, especially in the fundamental category of Type A innovation, just like what L. D. DeSimone, 3M's former CEO, once said, "Innovation is for every field." Innovation is still the mainstream, and global innovation is the source of 3M's competitive advantages. Specifically, in 3M's global business, there are seven distinct but intrinsically linked fields of innovative activities.

1) Selling and marketing technology

Many subsidiaries of 3M are basically just sales offices and lack new product development personnel and technological capabilities. In many regions, such as Asia, they do not really have a need to develop anything because the vast majority of the development work is done in the United States. However, in sales and marketing, there is still vast space for displaying their innovative capabilities. For example, for showing the customers 3M's capabilities to provide comprehensive solutions, in the United States, 3M invites customers to the 3M Park, but the subsidiary in the Philippines is more creative—they would send 30~40 employees to the site of an important customer to carry out a whole day of promotional display which the subsidiary in the Philippines refers to as the "salute to customers" activities. At 3M's subsidiary in R.O. Korea, the sales director of the department of commercial graphics developed an inductive method of sales promotion; the subsidiary invited large clients to Japan to see novel equipment that had not yet been widely used in R.O. Korea, so as to excite the competitive mentality in them. In addition, 3M's branch in the Philippines then adapted this method and once in a while took its important clients to the United States to see the new production equipment. Such innovations in sales and marketing techniques are done for the purpose of adapting to the local culture.

2) Product packaging

In the area of product packaging, there's great potential to tap into as follows: to make the colors and sizes of the packages attractive in the local markets, use a well-known and reputable local brand owned by a joint venture partner, sell several products as a set in one package, design product package in a way that ensures safe delivery, etc. Such approaches can be good for promoting sales. For example, packaging designs for 3M's U.S. products are made with the good highways and express transportation systems in the United States as the precondition. In order to adapt to the relatively bumpy roads in China, 3M's subsidiary in China has to improve the packaging of some products.

3) Product adaptation

Subsidiaries at different locations, by improving or customizing 3M's existing products, produce products that are in line with the local preferences. For example, 3M's scouring pad is an extension product of its nonwoven material technology platform. The sales of this product are poor in the Philippines. The subsidiary in the Philippines, after a survey of customer behaviors, found that many people are used to using coir for cleaning. Thus, they changed the color for 3M's scouring pad from green to coir-like yellow and the sales soared. The subsidiaries of 3M at various locations, by observing the local customs, have made some small improvements to their products based on such observations, which often result in big increases in their sales revenues.

4) Commercialization of new technologies

At 3M, the definition for innovation is "new idea plus behavior" that will bring about improvement or profit. The commercialization of new technologies is an indispensable part of innovative activities, and it is also the part that can actually bring profits. There is evidence that Japan is a fertile ground for the commercialization of new technologies.

5) Access to technological information

New technologies brought in from outside the company can make the fire of innovation rise higher. A big proportion of the world's PhD degree holders are in the various countries of the former Soviet Union; Russia and other regions of the former Soviet Union are rich in technological resources, and they form a vast repository of knowledge. However, there is a chance that those resources will remain unutilized. With the collapse of the Soviet Union, 3M has bought some important original technologies from the countries of the former Soviet Union. Such technologies have brought about a series of innovative activities and 3M has gained many patents as a result.

6) Cooperative design

From the very beginning, 3M has adopted a product design strategy that includes global input and multicultural cooperation and the company has been moving step by step toward achieving that goal. Since Sumitomo 3M in Japan has advanced technology, sophisticated laboratory operations, and an enormous domestic market, 3M and Japanese scientists have been at the forefront of joint design. For example, Japanese scientists participated in the initial design of the non-ozone-depleting chemical HFE and in the development of the brightness enhancement film.

7) Original invention

Various branches of 3M also have original inventions, including major technological breakthroughs and inventions that are based on existing products, but have brought about qualitative changes to them. A majority of the inventions fall into the latter category.

3. 3M's global innovation management

Those who visit 3M always try to take away a thing or two that can help them become more innovative, such as the "15% rule," but the most important thing is to foster an environment that makes innovation happen continually. Some particular innovation approaches do not necessarily apply in other industries. At 3M, innovation is the product of a complex environment; the most notable "endowment" of 3M is the cultivation of an overall work environment where a variety of factors interact and promote each other. Generally, there are three factors for nurturing innovation: culture, information exchange, and the elusive but crucially important management approach.

(1) 3M's innovative culture

3M's success is partly built on the company's cultural foundation of "innovation, creativity, and growth."

To talk about 3M's innovative culture, we must go back a hundred years and see how 3M started. In 1902, five entrepreneurs founded a company, purchased a piece of land that they believed to have corundum ore in it, and prepared to mine it. However, soon they discovered that the so-called mine contained only a low-grade calcium feldspar, not good enough even for abrasive, they later could only make sandpaper with it. It was not until 1907, when a young man named William McKnight joined the company, did 3M begin to truly develop a culture of systematic innovation. McKnight joined the newly established 3M as a bookkeeper right after his graduation and soon rose to become a leader of the company. In 1929, he became the president and since

1949 he served as the chairman of the board of directors until 1966. The 3M culture is established on the governing rules put forward by McKnight in 1948. McKnight is not only a manager but also a business philosopher. He created a corporate culture that encourages employee initiative, entrepreneurial spirit, and innovation; the many basic managerial policies that he founded are still 3M's guiding principles even today and will continue to play a vital role in the foreseeable future.

1) Tolerance, freedom, and security

A company cannot command people to be creative. What it can do is to try to create an atmosphere of tolerance, freedom, and security that is conducive to innovation and prosperity. Failures and mistakes are part of the innovative process, and tolerance of ineffectual ideas and learning from mistakes are important components of 3M's innovative culture. In 1902, the founding of 3M began with a mistake; a group of investors by mistake bought a mountain with no valuable ores and started to mine it for corundum—the company had not been making a profit for 14 years—and thus they spent a great deal of effort to create an atmosphere conducive to innovation. In 3M's innovative culture, failures and mistakes constitute important learning experience.

As business developed, there was even more need to entrust people with responsibilities and to encourage them to apply their initiative, which demanded a great deal of tolerance. As for those entrusted with rights and responsibilities, if they are good, they will want to work in their own way. It is just like what McKnight said: Mistakes will be made. But if a person is essentially right, the mistakes he or she makes are not as serious in the long run as the mistakes management will make if it undertakes to tell those in authority exactly how they must do their jobs. Management that is destructively critical when mistakes are made kills initiative. And it's essential that we have many people with initiative if we are to continue to grow.

2) Participation of the entire company

Some businesses will just appoint a new product manager and expect that he or she will be able to deliver new products rapidly. If that person has worked hard for 18 months but has hardly delivered anything, he or she would be removed from this duty or would leave the company. This process will be repeated with another newly appointed manager. However, innovation does not grow out of one or two individuals; it must be drawn from the entire organization and from every department of the company.

The unofficial motto of 3M is "every employee is a researcher." 3M has not isolated or protected any particular "scientific area" inside the company. Instead, whenever opportunities allow, it will let the company's system have the full play. For employees of every department, the company encourages them to become inventors or at least to be supportive of innovation. In order to promote innovation, the company stipulates that 30% of the annual sales revenue must come from products developed in the previous four years and that 10% of the annual sales revenue must be from products developed within a year.

3) 15% rule

An important reflection of 3M's support for innovation is in the 15% rule. The 15% rule originally referred to the practice that 3M's laboratory personnel are allowed to freely carry out research other than their mandated tasks during the coffee break, lunch break, and spare time. The 15% is not really a strictly fixed ratio and not everyone uses it, but some people use more than 15% of their time. The results of the 15% rule are many new business areas important for the company.

4) Well-intended ignoring and threat

Management should know when not to manage; they should allow the various innovative factors to play out and should not interfere. This is also an important lesson that 3M has learned. The 15% rule is a typical example of well-intended ignoring. 3M managers are obligated to turn a blind eye on employees engaging in their own research during work hours, so long as they finish their assigned tasks, even if a program exceeds its deadline or a very promising project fails in the end. In a delicate stage of product development, managers may act to protect the research program and put forward a laboratory version of the "15% rule."

3M managers will also threaten to cancel a research program so as to pressure researchers into achieving breakthroughs. Masking tapes, Post-it notes, nonwoven technology, immune response modifiers, and so on, all have once been under the threat of cancellation.

5) Technology belongs to the company

Products fall within the responsibilities of various divisions, and each division directly targets certain markets and responds to its customers. However, technology is shared across the entire company and belongs to the company as a whole. Every division can use the entire company's technological resources, and it also has a responsibility to make available the technical requirements of its customers for sharing within the entire company. 3M also promotes the interconnection of its core technologies so as to take advantage of the strength from collaboration and create new products and new applications.

6) Tradition and innovation

Many companies are willing to hire short-term contract workers in the mistaken belief that new employees in the knowledge management system will produce innovative ideas. However, such short-term employees will be lacking in loyalty to the company and will not be interested in sharing their knowledge for the success of long-term innovation. They probably do not expect the freedom to innovate; they do not look for opportunities to help others so as to be helped by others. They know that they will leave in the short term, and thus, their primary interest lies in the professional experience to be gained; they work to ensure that their resumes will interest the next employer.

Innovation often occurs at the edges, and it does not always start with a "eureka!" scream. It may be a mild passion that needs to slowly unfold. However, if the work is for short term, who will have the time for innovation?

3M is different. 3M encourages a sense of tradition; lifelong employment and promotion from within are important 3M policies. Employees work at 3M for decades and their turnover is very low. 3M employees enjoy continued employment and a global employee network, which lead them to help others time and again without expecting reciprocatory favors. However, these people know for sure that, over the next 10 years or 20 years, when they need help, they will get it. Canadian biologist Hans Selye, PhD, uses the expression "altruistic self-centered" to describe those who help others for their own benefits. Only continuity can allow such an atmosphere to flourish.

(2) Internal communication and exchange

3M is truly a hundred-year-old time-honored brand. In most people's perception, the big companies that were established at the beginning of the 20th century are still operating successfully, such as DuPont and Siemens. However, in fact, according to a 1998 study by Hannah, the more likely outcome for a one-time large company is that it has failed in competition: of the top 100 largest companies around the globe, only 20 that were leaders in early 20th century were still among the top 100 largest companies in 1995. The United States has a fiercely competitive market, a robust entrepreneurial spirit, and an innovative tradition that is more relentless than "creative destruction." Of the U.S. companies that were among the top 100 toward the end of the 20th century, only 26% are still among the top 100 at the beginning of the 21st century; however, for the United Kingdom and Germany, the figures are 75% and 66%, respectively, and 3M is among the 26%.

Regardless of the size of a company, an important problem is that one group or department does not know what the other groups or departments are doing; this is especially true in large companies, whose heavy structures, cumbersome internal guidelines, rigid procedures, and entrenched bureaucracy, all make the communication between each department and its employees more difficult, thereby inevitably leading to a lack or slowdown of technology transfer and innovation as well as the loss of opportunities. Successful large companies may produce organizational mechanisms that negatively affect innovation and responses to customer demands; they become the victims of their own success.

3M has about 90,000 employees. Of the 90,000 employees, more than 35,000 work in subsidiaries outside the United States and most of such employees are localized—around the globe only a few hundred employees are foreigners working outside their own countries. Employees from multiple cultures and multiple backgrounds have made communication a necessary part of 3M's success. For smooth communication among various technical personnel in marketing, sales, and manufacturing and customers, 3M encourages formal and information networking activities that can ensure coordinated actions.

1) Infrastructure helpful for communication

3M adopts the best technologies and methods to enhance communication and interaction among teams. Formal training programs, at-work learning, support department, intranet, the Internet, Lotus Notes (groupware released by Lotus Software, the United States), videoconferencing, IT-based database, and so on are available for a wide range of employees. Technology planning and coordinating groups are responsible for updating and maintaining best practices and mission-critical databases. With over two dozens of its core technologies, the company knows who is working on what subject.

To better facilitate communication among people of different divisions and different functional departments, 3M extended their innovative talent to the field of internal communication. In Europe, many of 3M's subsidiaries, after decades of efforts, have built up local operations of enormous scale. However, for many branch companies, to keep the websites of a bunch of different languages up-to-date, the costs and heavy workload are unbearable. The webmaster of 3M's Europe headquarters started a dynamic online publishing method, whereby 3M employees can input information in their local languages into databases, which will then be used to generate real-time web pages. In this way, website development work falls directly into the hands of local employees and such employees can easily update website content at a low cost and meet the needs of particular countries. When such dynamic web publishing pages first opened, there were only two hundred static web pages; one year later, there were already more than two thousand web pages. The costs for installing such a system have been recovered within six months. This project has provided a template for employees and clients in various regions to share information generated in different languages, and it has expanded website workflow to the development of other systems, making speedy management approval and news materials being released a lot easier. Dynamic online publishing has proved to be so successful in Europe that there have been a groundswell of support for its deployment companywide.

2) Tacit-to-tacit knowledge management

Japanese scholar Nonaka believes that knowledge within an organization can be achieved through the following four conversion processes.

Socialization—sharing of tacit knowledge among individuals, from tacit knowledge to tacit knowledge of an even deeper level;

Externalization—explaining tacit knowledge and converting it into explicit knowledge;

Explicitation—diffusing written knowledge among the members

of an organization and going from explicit knowledge to complex and synthesized explicit knowledge;

Internalization—moving from explicit knowledge to tacit knowledge of the organization.

In these four areas of knowledge management, 3M places much emphasis on "tacit-to-tacit" knowledge transfer, encouraging people to acquire and share information and to convert information into knowledge. This means that people need to spend a certain amount of time together and to get knowledge through psychological closeness. In many companies, face-to-face meetings have been reduced in order to lower expenses, but at 3M, 3M is said to represent "Meetings, More Meetings," which is the proof of 3M's effort to ensure the efficiency of scattered teams.

3) Sharing of intellectual assets and cooperation

The products of 3M run across many different markets. In order to develop products suitable for different and independent markets, 3M's technology people include scientists, PhDs in medicine, mechanical engineers, optical physicists, computer programmers, statisticians, and so on. Along with such diversified technology, people are in diversified horizontal organizations such as business divisions and departments. Within a business division or department, there are often employees or teams working for different products, research studies, designs, and so on. In addition, these teams may be far away from the company's headquarters; such situation will lead to segregation among teams or employees, hindering the rapid diffusion of knowledge.

At 3M, great importance is attached to cross-divisional and cross-border cooperation, allowing employees to share their intellectual assets within the scope of their interest. People from different backgrounds gather face to face or through videoconferencing technologies within an environment of supportive comments. Many new ideas can come out of this. The 3M Tech Forum is an organization that belongs to technology personnel where they can share ideas about technologies, policies, and programs. In addition to lectures and problem-solving meetings, the 3M Tech Forum is the initiator for an annual event where various branches display their best products and technologies and is also the sponsor for specific activities in various technological fields (the main sponsor). The company allows special technology interest groups to organize issue discussions during work hours across various departments.

4) Continuity

3M's tradition is also conducive to communication.

A system of lifetime employment and promotion from within means that the 3M people know well about each other; senior members have already visited or worked in various countries or are currently doing so. They have extensive personal contact with various employees, and such contact leads to knowledge transfer, which effectively crosses cultural boundaries. Storytelling is even more of a tradition built upon company history. The members of top management who joined the company when they were young and have absorbed the tradition and stories of the company retell those stories and thereby promote the company's value system and atmosphere of innovation.

(3) Management approach that supports global innovation

1) Organizational structure

Considering the organizational structure, 3M's international business division is placed under the charge of one of the few top leaders, executive vice president of the company. This also shows the importance of the business operations outside the United States. The overall international business operations cover four regions (see Figure 5); an executive officer who is responsible for various regions and a managing director who is responsible for the operations of individual countries report to the executive vice president of the company. Every region and each major subsidiary have a technology director, a production director, and a marketing director. It takes on a mesh structure with many intersecting points, which is balanced through 3M's 40 plus business divisions, and each of these divisions is managed as an independent company within 3M. Such small decentralized and autonomous business units allow them to reduce bureaucracy to the lowest level and to focus on new ideas and their own customer bases.

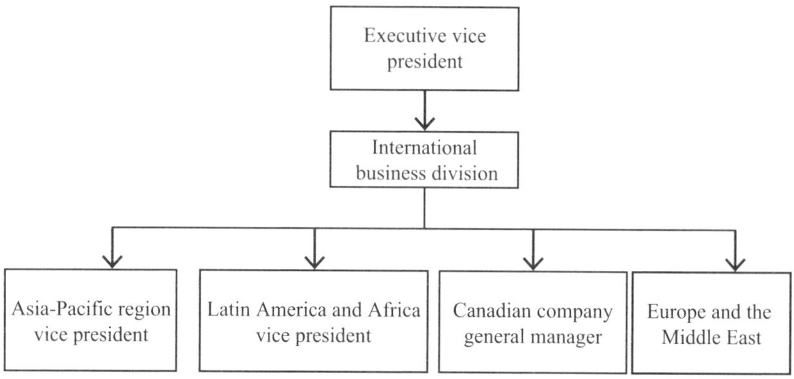

Figure 5 3M's international business division

2) Full-range overseas presence

The employees in control of 3M's overseas subsidiaries are mostly dispatched from the company's headquarters, totaling about 180. When compared to the 30,000 employees of 3M's overseas subsidiaries, the number of overseas managers is very small, but such managers are on pivotal positions in technological interaction. Among them, managing directors of various countries traditionally are people dispatched from the United States. However, in recent years, there have been more and more cases where someone was sent over from one non-U.S. country to another to take on such a position, such as from Japan to Thailand, from the United Kingdom to South Africa, and so on. Such cases also help to spread technology and employee expertise in those countries. The technology directors of various countries are often sent over from the United States. Before being dispatched, they generally work in the research institutes in Saint Paul. For the technology transfers between the United States and other countries, such technology directors can serve as an ideal communication channel.

In addition, in order to spread the experience, insights, and innovative ideas of sales staff and scientists from Saint Paul to other places around the world, 3M implements a global mentoring program. 3M's corporate scientists are regularly sent to overseas laboratories, where they stay for quite some time and hold technology seminars to have enough influence on promising young employees there. All corporate scientists of 3M have a long list of various inventions and patents as well as records of success in commercializing products, which command respect in young researchers and also produce inspirational effect on them. For example, in China, 3M implements a "Triple Three" program, whereby three U.S. sales representatives are sent to China to tutor their Chinese counterparts for a period of three weeks three times a year.

Selecting employees in other countries and sending them to the United States is another important component of 3M's overseas exchange strategy. Large numbers of employees come from other countries to the United States to learn particular technologies or for short-term visits, thereby establishing contact with 3M's U.S. technology institutions. When working in the United States, these people can participate in activities, such as product improvement, cooperative design, and so on; familiarize themselves with the enterprise culture; and become key contact people and coordinators for cross-border technology exchanges and innovation after returning to their own countries.

3) Community of Enterprise Development and Transnational Team

The Community of Enterprise Development is a group set up in 1997. Its purpose is to foster new business within 3M, and the group invests two-thirds of its energy in regions outside the United States. Most of the new businesses supported by the Community of Enterprise Development are proposed by the employees. After a proposal passes the initial review, the Community of Enterprise Development will provide funding for the development of the business plan and will introduce it into an existing business unit or into a business unit temporarily set up with funding from the Community of Enterprise Development. The Community of Enterprise Development will also monitor companies that have just emerged in related markets or technological fields in an effort to keep 3M in line with the new trends.

Transnational Team grew out of Europe. The group was first named European Management Action Team (EMAT), and its purpose is to make headway in markets with major opportunities by concentrating on more energy, better cooperation, more rapid pace, and a spirit of systematic effort and perseverance. They regularly gather relevant personnel from the United States and in local area to share information and make decisions. In encouraging the U.S. members to spend more time and energy on markets outside the United States, this group played an important role; meanwhile, regular group meetings have allowed non-U.S. members to be able to express their views more forcefully and to shoulder more important responsibilities in relation to customers. The success of the European team prompted 3M to replicate the model in other regions and produced Japanese Management Action Team, Asia-Pacific Management Action Team, and so on. 3M closely observes the performance of these groups and spreads successful multinational cooperation activities to other fields of the company.

4) Recruiting and retaining the right employees

"The core of 3M are people. Whether or not we will succeed depends on whether or not our employees have the ability to succeed." (McKnight)

The employees that 3M finds ideal are those who want to start something new and not those who just want to continue with the traditional business. They are more interested in the freedom to do their own things rather than money and power. They are action-oriented and are ready to take risks for the future. 3M recruiters look for problem solvers who are creative (those who can live in contradictions and, at the same time, regard opposite things as real), have strong ethical values, and have a broad interest (those who have a broad interest are eager to learn, are willing to explore ideas with others, adopt a multidisciplinary approach, and are happy to establish a communication network through both face-to-face interactions and IT tools).

Informal lifetime employment and lower employee turnover can nurture the spirit of adventure, which will bring employees to absorb more experience through their professional careers.

5) Incentives and recognition

3M knows that the inventors of new industries and new products are the company's key figures and that uncovering the commercial value of a new product for the company is everlasting. Enterprises have a variety of procedures and methods to encourage and commend innovation; similarly, 3M has a procedure and its employees know more about "innovation heroes" rather than senior managers. Ensuring inventors being properly recognized can also stimulate employees to act. In addition to the regular tools for stimulating creativity and innovation, such as salary increase and promotion, 3M developed the following series of programs and incentives to motivate and recognize its most valuable employees.

Exhibition: This exhibition consists of technology exhibits that are good for product development. 3M members are invited from around the company, using the principle of confidentiality, to test whether certain technology can be applied in a different field of the company's business.

Super-priority Development Program: This program is used to provide development funds for projects with the most potential and the largest market.

Creativity Funds: Technology personnel can apply for creativity funds to purchase equipment and develop their "15%" ideas. When they spend their own time to develop their 15% projects, they can also use the grant to pay the temporary laboratory for doing the current work.

Alpha Funds: They are grants for developing innovation in management, marketing, and other non-technology fields, such as new technological process.

Golden Step: It is an award for any team whose new product—within three years of its introduction—has turned profitable and the sales revenue has reached $5 million.

Pioneer Program: It rewards a team of sales, marketing, logistics, finance, and production staff members who work on developing and introducing new product in a country.

Commercial Pioneer Award: It is an award for recognizing a team that has come up with a new idea in another country and then launched it in their own country.

Inventor Award: It is an award for recognizing employees who have successfully developed a new product and new technology through indirect research.

Carlton Society: It is an honor organization for those who have made extraordinary contributions to 3M in S&T.

6) Innovation-inducing informal method: friendship and volunteer action

The global presence strategy that has been in practice at 3M for years, coupled with the very low overall employee turnover, has created a globally interconnected friendship network that crisscrosses the official organizational structure and complements its functions. The result of the American Midwest hospitality is that, at 3M, there is still this warmth between people that you can feel everywhere.

For innovative activities, friendship brings with it an unseen but valuable support; good interpersonal relationships can provide an indirect driving force for transnational innovation activities. If a future innovator at 3M should need information about how to solve a problem, as long as one of its 70,000 employees could solve the problem before that, good interpersonal relationships and friendships would reduce the trouble of long-term background search. If a conflict happens between two members over a particular project or something else, friendship can help them to find a reasonable compromise.

By strengthening interpersonal relationships and putting them into practice, volunteer actions solidify friendships.

3M's language community. It all started more than 30 years ago, when a group of German employees were stationed in Saint Paul and they often gathered and talked with each other in their mother tongue. As the gathering got larger and larger in size and became more and more formal, some others expressed their interest in learning German. This is how the language community was founded and gradually developed.

3M's volunteer program for employees serving overseas. The program originated from a mass campaign launched by the spouses of employees stationed abroad, who were hoping that 3M would play a more proactive role in supporting the employees and their families that were stationed overseas. The activities of the program were initially designed for helping employees serving overseas and their families to adjust and adapt to their life abroad, but now it appears that it also benefits 3M a lot because it helps employees working overseas in an unfamiliar environment to establish contact with the right people and to have the opportunity to obtain the hard-won knowledge of their predecessors.

(4) Globally balanced management

Global innovation activities have brought new creative opportunities to 3M, but they have also brought challenges in terms of management: 3M's full line of products are very extensive—which pose a great deal of difficulties for newly established subsidiaries; at the same time, there are various needs of 3M's various departments to be met; as a result, some employees are busy working on things related to the U.S. headquarters or other regional headquarters and do not have time or energy to work with customers; the existing products and procedures are convenient to every employee, but they pose the danger of inhibiting innovation activities; and as 3M builds local production facilities worldwide, questions about the pace of investment and about global efficiency have always come as a headache. Most experienced managers of 3M know how to strike the right balance between control and support. The balancing factors by which 3M promotes vigorous development of global innovation activities are shown in Figure 6.

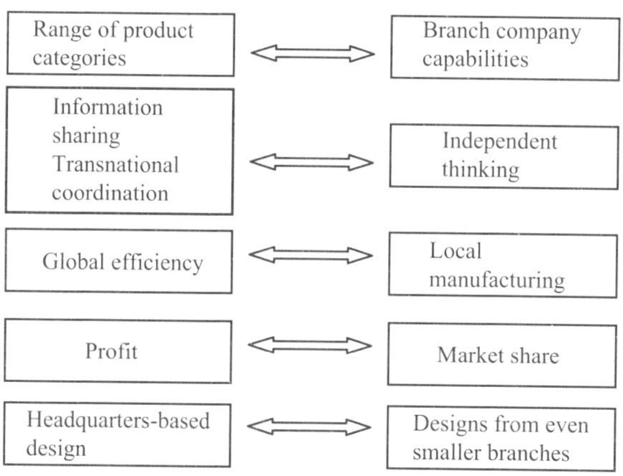

Figure 6 Key factors of balanced global management

References

Bartlett, C. A. & Ghoshal, S. Managing innovation in the transnational corporation. In: Bartlett, C. A., Doz, Y. & Hedlund, G. (Eds.). *Managing the Global Firm*. London: Routledge, 1990.

Birr, K. *Pioneering in Industrial Research*. Washington, D.C.: Public Affair Press, 1957.

Brand, A. Knowledge management and innovation at 3M. *Journal of Knowledge Management*, 1998, 2(1): 17-22.

Brockhoff, K. *Internationalization of Research and Development*. Berlin: Springer-Verlag, 1998.

Cantwell, J. A. The globalization of technology: what remains of the product cycle model? *Cambridge Journal of Economics*, 1995, 19(1): 155-174.

Cantwell, J. A. & Hodson, C. Global R&D and UK competitiveness. In: Casson, M. (ed.). *Global Research Strategy and International Competitiveness*. Oxford: Basil Blackwell, 1991.

Cantwell, J. A. & Piscitello, L. Accumulating technological competence: its changing impact on corporate diversification and internationalization. *Industrial and Corporate Change*, 2000, 9(1): 21-51.

Casson, M. & Singh, S. Corporate R&D strategy: the influence of firm, industry and country factors on the decentralization of R&D. *R&D Management*, 1993, 23(2): 91-107.

Chen, J. *Sustainable Development—New Perspective on Enterprise Technology Innovation*. Beijing: Science Press, 2001.

Dalton, D. H. & Serapio, M. G. *Globalizing Industrial Research and Development*. Washington, DC: U.S. Department of Commerce, Technology Administration Office of Technology Policy, 1999.

Gassmann, O. & Von Zedtwitz, M. New concepts and trends in international R&D organization. *Research Policy*, 1999, 28(2-3): 231-250.

Gerybadze, A. & Reger, G. Globalization of R&D: recent changes in the management of innovation in transnational corporations. *Research Policy*, 1999, 28(2-3): 251-274.

Granstrand, O., Hakanson, L. & Sjolander, S. *Technology*

Management and International Business: Internationalization of R&D and Technology. Chichester: Wiley, 1992.

Gundling, E. *Fertile Soil for Innovation—3M's Innovation Mechanism*. Chen X. S., Chi, J. C. & Zhang, H., trans. Beijing: Huaxia Press, 2001.

Hakanson, L. Locational determinants of foreign R&D in Swedish multinationals. In: Granstrand, O., Hakanson, L. & Skolander, S. (eds.). *Technology Management and International Business*. Chichester: Wiley, 1992.

Hannah, L. Survival and size mobility among the world's largest 100 industrial corporations, 1912–1995. *American Economic Review*, 1998, 88 (2): 62-65.

Hawkins, L. A. *Adventure into the Unknown*. New York: William Morrow & Co., 1950.

Jiang , J. H. Successful conversion of science and technology—GE of the United States at a glance. *East China Science and Technology*, 1996(6): 7-8.

Kanter, R. M. *Innovation: Breakthrough Ideas at 3M, Dupont, GE, Pfizer and Rubbermaid*. New York: Harper Business, 1997.

Kumar, N. Determinants of location of overseas R&D activity of multinational enterprises: the case of US and Japanese corporations. *Research Policy*, 2001, 30(1):159-174.

Li, H. J. *New Development of TNCs and Analysis of the Economic Effect*. Harbin: Heilongjiang People's Press, 2002.

Li, M. C. The changes in the development approach of GE institute. *Western High and New Technology*, 1997(4): 31-33.

Lin, J. C. & Chai, Z. H. Analysis on the trend of internationalization of technological research and development in transnational corporations. *World Economy*, 1998(7): 35-39.

Liu, L. History and current state of R&D at GE of the United States. *Science, Technology, and Dialectics*, 1998(5): 44-48.

Liu, L. *Plug in the Wings of Science and Technology: The Rise of Germany's Chemical Industry*. Taiyuan: Shanxi Education Press, 1999.

Liu, L. *GE—Harvard in World's Business Community*.

Shijiazhuang: Hebei University Press, 2001.

Lowe, J. *Jack Welch Speaks*. Haikou: Hainan Press, 1999.

Luo, H. The triangular structure that created the GE miracle. *Enterprise Reform and Management*, 2002(2): 40.

Mees, C. E. K. *The Organization of Industrial Scientific Research*. New York: McGraw-Hill Co., 1960.

Nonaka, I. & Takeuchi, H. *The Knowledge-creating Company: How Japanese Companies Create the Dynamics of Innovation*. Oxford: Oxford University Press, 1995.

Pearce, R. D. Decentralized R&D and strategic competitiveness: globalized approaches to generation and use of technology in multinational enterprises (MNEs). *Research Policy*, 1999(28): 157-178.

Pearce, R. D. Factors influencing the internationalization of R&D in multinational enterprises. In: Buckley, P. J. & Casson, M. (eds.). *Multinational Enterprises in the World Economy*. London: Elgar, 1992.

Pearce, R. D. & Singh, S. *Globalizing Research and Development*. London: Macmillan, 1992.

Reich, L. S. *The Making of American Industrial Research*. Cambridge: Cambridge University Press, 1985.

Rosenbloom R. S. & Spencer, W. J. *Engines of Innovation: U.S. Industrial Research*. New York: Columbia University Press, 1983.

Slate, R. *Get Better or Get Beaten! 31 Leadership Secrets from GE's Jack Welch*. Beijing: Mechanical Industry Press, 1999.

Slate, R. *Jack Welch and the G.E.Way*. Beijing: Mechanical Industry Press, 1999.

Tichy, N. *Control Destiny: GE's Reform Process*. Shanghai: Shanghai Translation Publishing House, 1996.

Unknown. Ultimate globalization—an interview with president of GE China, Wang Jianmin. *China Market*, 2001(5): 48.

Wang, X. C. Analysis of policies and directions of international cooperation in science and technology. *Science of Sciences and Science and Technology Management*, 1997(4): 27-29.

Wise, G. & Whitney, W. *General Electric and the Rise of U.S. Industrial Research*. New York: Columbia University Press, 1983.

Xu, Q. R. *Research, Development, and Technology Innovation Management*. Beijing: Higher Education Press, 2000.

Zhang, D. K. Dilemma of succession: "same" or "different"?. *Chinese and Foreign Management*, 2001(9): 42-43.

Zhang, Z. C. *Challenging the Limit—Demystifying the GE Miracle.* Guangzhou: Guangzhou Press, 2001.

Appendix 2

Appendix 2

Questionnaire for China's pioneer enterprises in R&D internationalization

Notes on completing the questionnaire:

Please fill out the questionnaire with the current situation of your company. Mark "√" in the cell of your choice. The meanings for score value from 1 to 5 are as follows: 1 means either that this point is the least important or that it differs from the actual situation in your company the most; 2 means not very important; 3 means medium; 4 means important; 5 means that this point is the most important or that it fits the actual situation in your company the best.

Your department:________ Your position: _______

1. Basic situation of the company and its performance in overseas R&D

(1) Basic situation of the company

(a) Company name:__________

(b) Location of headquarters:__________

(c) Year in which company was established:_________

(d) Leading products:__________

(e) Core technology:__________

(f) Number of employees:__________

(g) Number of R&D personnel:__________

(h) Number of international R&D personnel:__________

(i) Total assets:__________

(j) Total net assets:__________

(k) Sales revenue:__________

(2) Evaluation of performance in international R&D activities

(a) Success rate of overseas R&D projects is roughly__________%

(b) This year's proportion of sales revenue accounted for by new products developed by overseas R&D activities since such activities started is__________%

(c) Compared with the time needed when there were no internationalized R&D activities, how much faster can a new product get to the market:__________%

(d) Compared with the time needed when there were no internationalized R&D activities, how much faster can a new product turn profitable:__________%

(e) Compared with the time when there were no internationalized R&D activities, how much less is the company's waste in R&D spending:__________%

(3) Please fill out the level of importance or of success for the following items:

No.	Item	Score				
		1	2	3	4	5
1	Contributions to your company's total profit from overseas R&D projects					
2	Degree of success for overseas R&D activities, relative to the various resource inputs					
3	Profitability which is the gains from your company's technological innovation relative to its costs, since the start of your company's overseas R&D activities					

2. Strategic factors for your company in implementing overseas R&D

(4) What is the current situation of your company's R&D internationalization? What is the trend for the next five years?

No.	Item	Score									
		Current					Next five years				
		1	2	3	4	5	1	2	3	4	5
1	Introduction of foreign technology through licensing										
2	Recruitment of foreign scientists and technological personnel										
3	Acquisition of overseas enterprises through M&As (mainly for R&D reasons)										
4	Overseas patent applications by overseas subsidiaries										
5	Establishment of cooperative relationships in research with overseas universities and research institutes										
6	Establishment of technological alliances with other companies overseas										
7	Direct investment abroad and establishment of R&D branches (including R&D branches set up upon existing overseas subsidiaries, such as technological testing points and so on)										
8	Research on technologies outside the main business										

(5) What are the driving factors of your company's R&D internationalization? How important are these factors? How will their importance probably change in the next five years?

No.	Driving factor	Score									
		Current					Next five years				
		1	2	3	4	5	1	2	3	4	5
1	Responding to competition and implementing development strategy										
2	Expanding and enhancing potential technological capabilities										
3	Providing necessary technology support for local subsidiaries (especially technology transfer from the domestic to the local market and technology support in product and process and so on)										
4	Introducing new products to the local market as soon as possible										
5	Making use of good local R&D infrastructure										
6	An inevitable trend after internationalization of the company's production and operations										
7	Better adapting to the special needs of consumers in different parts of the world and the needs for producing specialty products										
8	Needs of local subsidiaries for development and marketing										
9	Local preferential policies (such as taxes, low-interest loans, etc.)										
10	Needs for setting up technology monitoring points, understanding competitors, and determining the trend of technological development										
11	Recruiting outstanding R&D personnel and absorbing advanced technology and scientific knowledge from abroad										
12	Local restrictive policies (referring to those conditioning entry into the local market on the establishment of R&D institution and those restricting imports and limiting profits that flow back to the home country and so on)										

(Continued)

No.	Driving factor	Score									
		Current					Next five years				
		1	2	3	4	5	1	2	3	4	5
13	Ease as a result of the development in telecommunications and information technology										
14	Cooperation with local R&D projects										
15	Getting close to suppliers										

(6) How many R&D branches has your company set up in the following overseas regions? How is the number of R&D branches expected to change in the next five years?

No.	Region	Score									
		Current					Next five years				
		0	1–2	3–5	6–8	> 8	0	1–2	3–5	6–8	> 8
1	United States										
2	Japan										
3	Europe										
4	Southeast Asia										
5	Other regions										

(7) What is the earliest time your company started to set up R&D branches in the following regions (if your company has not yet set up any R&D branches overseas, leave the following table blank)?

No.	Region	Time
1	United States	
2	Japan	
3	Europe	
4	Southeast Asia	
5	Other regions	

(8) About the following inhibiting factors of internationalization, how important are they? How will their importance change in the next five years?

No.	Inhibiting factor	Score									
		Current					Next five years				
		1	2	3	4	5	1	2	3	4	5
1	Need for supervision and control of R&D										
2	Considerations for avoiding risks of leaking core technologies										
3	Need for closely linking R&D to the domestic market										
4	Considerations for optimizing the scale of R&D										
5	Reducing the costs of coordination and communications										
6	Spatial as well as language and cultural barriers										
7	Considerations for avoiding repetitive R&D constructions										
8	Good domestic R&D infrastructure and close links with domestic universities and research institutes, and thus lack of motivation for carrying out overseas R&D activities										
9	Policy factors (referring to host country policies)										

(9) What are the idea sources of your company's overseas R&D? How important are these sources? How will their importance change in the next five years?

No.	Source of R&D ideas	Score									
		Current					Next five years				
		1	2	3	4	5	1	2	3	4	5
1	Customers at the location of the overseas R&D units										
2	Marketing departments overseas										
3	Production departments overseas										

(Continued)

No.	Source of R&D ideas	Score									
		Current					Next five years				
		1	2	3	4	5	1	2	3	4	5
4	Cooperation with universities and research institutes overseas										
5	R&D departments of the parent company										
6	Other departments of the parent company										
7	Competitors overseas										
8	Various periodicals, plans by overseas local governments, etc.										
9	Local enterprises acquired through M&As										

(10) What are the organizational types of your company's overseas R&D branches? How will this change in the next five years?

No.	Type	Score									
		Current					Next five years				
		1	2	3	4	5	1	2	3	4	5
1	Basic scientific research (long term)										
2	Applied research that is based on basic research, creating possible commercial applications (medium term)										
3	Commercial product development for the market (short term)										
4	Improvement of product development (short term)										
5	Technological process transformation for the market (short term)										

3. Funding and resources

(11) Please fill in the actual percentage of how R&D funds in your company are used from year ____ to ____ and predict changes in the future.

No.	Item	Past	Expected changes
1	Total R&D expenditure (both domestic and overseas)/total sales revenue		
2	Total overseas sales revenue (including company export and overseas sales by overseas subsidiaries)/total sales revenue		
3	Total overseas sales revenue of overseas subsidiaries/total sales revenue		
4	Overseas R&D expenditure/total R&D expenditure		
5	Overseas expenditure on basic research/total R&D expenditure		
6	Overseas expenditure on applied research/total R&D expenditure		
7	Overseas expenditure on experimental development/total R&D expenditure		
8	Overseas expenditure on basic research/total expenditure on basic research		
9	Overseas expenditure on applied research/total expenditure on applied research		
10	Overseas expenditure on experimental development/total expenditure on experimental development		

(12) What is the total amount that your company spent on overseas R&D in year ____? What is the total amount that this spending is expected to reach in the future?

Year	Amount/Yuan						
	0–500	>500 ~2000	>2000 ~5000	>5000 ~10000	>10000 ~25000	>25000 ~50000	>50000
____							
Future							

(13) How are the R&D funds for your company's overseas R&D branches distributed among the various funding sources?

No.	Funding source	Proportion				
		<10%	>10% ~20%	>20% ~30%	>30% ~40%	>40% ~50%
1	R&D funding provided by parent company					
2	R&D funding provided by overseas subsidiaries					
3	R&D funding provided by Chinese government					
4	R&D funding provided by overseas cooperative and joint venture companies					
5	R&D funding provided by various domestic NGOs					
6	R&D funding provided by various overseas NGOs					

No.	Funding source	Proportion				
		>50% ~60%	>60% ~70%	>70% ~80%	>80% ~90%	>90% ~100%
1	R&D funding provided by parent company					
2	R&D funding provided by overseas subsidiaries					
3	R&D funding provided by Chinese government					
4	R&D funding provided by overseas cooperative and joint venture companies					
5	R&D funding provided by various domestic NGOs					
6	R&D funding provided by various overseas NGOs					

(14) With regard to technological reserves and capability accumulation for overseas R&D, how important are the following factors?

No.	Factor	Score				
		1	2	3	4	5
1	Understanding of the dynamics of technological development (especially the dynamics of international technological development)					
2	Pre-research on future technology, with input of both human and material resources					
3	Establishing learning-oriented international R&D organizations and accumulating experience continuously					
4	Ability to accumulate experience from failed overseas R&D projects					
5	Ability to build up sound technical documentation for overseas R&D projects					
6	Attention to the cultivation of international researchers					
7	Reflection in the technological strategy of fostering technological capabilities through R&D internationalization					

4. Personnel selection and organizational structure

(15) What is the current organizational form of your company's overseas R&D branches and in the next five years?

No.	Organizational form	Score									
		Current					Next five years				
		1	2	3	4	5	1	2	3	4	5
1	R&D activities are carried out mainly in the home country, and little attention is paid to overseas R&D activities										
2	R&D headquarters is set up in the home country, and this headquarters directs overseas R&D activities and exerts strong control over overseas R&D activities										
3	Regional R&D centers are set up in parallel in a number of regions of the world, and such centers guide their subordinates to carry out R&D activities										
4	Independent R&D institutions are set up worldwide, and the headquarters exerts relatively weak control over them										

(Continued)

No.	Organizational form	Score									
		Current					Next five years				
		1	2	3	4	5	1	2	3	4	5
5	A closely interconnected worldwide R&D network is constructed, and this network enhances the exchanges as well as communications and the information transfer among the R&D institutions that are dispersed worldwide										

(16) Currently, what is the proportion of foreign researchers in your company's overseas R&D branches in different regions and what is the future trend of this proportion?

Item	Proportion				
	<10%	>10% ~20%	>20% ~30%	>30% ~40%	>40% ~50%
United States					
Japan					
Europe					
Southeast Asia					
Other regions					
Current					
Future					

Item	Proportion				
	>50% ~60%	>60% ~70%	>70% ~80%	>80% ~90%	>90% ~100%
United States					
Japan					
Europe					
Southeast Asia					
Other regions					
Current					
Future					

(17) What is the current proportion of the foreign managers in your company's overseas branches and what is the future trend of this proportion?

Time	<10%	>10% ~20%	>20% ~30%	>30% ~40%	>40% ~50%
Current					
Future					
Time	**>50% ~60%**	**>60% ~70%**	**>70% ~80%**	**>80% ~90%**	**>90% ~100%**
Current					
Future					

(18) What is the current proportion of the foreign project managers for your company's ongoing overseas R&D projects and what is the future trend of this proportion?

Time	<10%	>10% ~20%	>20% ~30%	>30% ~40%	>40% ~50%
Current					
Future					
Time	**>50% ~60%**	**>60% ~70%**	**>70% ~80%**	**>80% ~90%**	**>90% ~100%**
Current					
Future					

(19) How important are the following factors for selecting managers of overseas R&D branches?

No.	Factor	Score				
		1	2	3	4	5
1	High technological level and broad knowledge					
2	Excellent skills in communication and coordination with researchers of various nationalities					
3	Familiarity with local social and cultural background					
4	Skills in exchange and communication with external people to the organization and in gaining access to extensive knowledge					

(Continued)

No.	Factor	Score				
		1	2	3	4	5
5	Skills in exchange and communication with the R&D headquarters of the parent company and in timely reporting to the headquarters about the situation in the branch and relaying various messages from the headquarters to the R&D branch					
6	Innovative spirit and ability to stimulate enthusiasm and creativity of the employees					
7	Full understanding of the parent company's technology strategy and even participation in the formulation and implementation of such strategy					
8	Ease of finding					
9	Generally between 40 and 50 years of age					

(20) How important are the following factors for selecting researchers of overseas R&D branches?

No.	Factor	Score				
		1	2	3	4	5
1	Researchers' technological ability					
2	Researchers' ability to learn (including Chinese and foreign researchers)					
3	Researchers' creativity (including Chinese and foreign researchers)					
4	Researchers' ability to gather information (including Chinese and foreign researchers)					
5	Researchers' insight (including Chinese and foreign researchers)					
6	Researchers' ability to communicate, especially with researchers of various nationalities (including Chinese and foreign researchers)					
7	Researchers' professional dedication (including Chinese and foreign researchers)					
8	Researchers' ability for international cooperation and exchange (including Chinese and foreign researchers)					
9	Researchers' language skills (including Chinese and foreign researchers)					
10	Foreign researchers' understanding of Chinese cultural and social customs					
11	Chinese researchers' understanding of and familiarity with host country's social and cultural background					

(21) How are decisions about R&D projects in your company's overseas R&D branches made?

No.	Factor	Score									
		Current					Next five years				
		1	2	3	4	5	1	2	3	4	5
1	Branches determine R&D projects and independently implement the development										
2	Headquarters works on project design and dispatches projects to various overseas R&D units										
3	Branch units work on project design and submit major projects to headquarters for approval										
4	Branch units work on project design and submit projects to headquarters for approval										

5. Exchange and communication

(22) What is the current situation of exchange and communication between the parent company and the company's overseas R&D branches?

No.	Item	Score				
		1	2	3	4	5
1	Parent company and its overseas R&D branches mainly communicate via face-to-face contact					
2	Parent company and its overseas R&D branches mainly communicate through IT tools					
3	Overseas R&D branches mainly communicate with each other via face-to-face contact					
4	Overseas R&D branches mainly communicate with each other through IT tools					
5	Degree of your satisfaction with communication between the parent company and the overseas R&D branches					
6	Degree of your satisfaction with communication among the overseas R&D branches					

(23) How important are the following means of communication?

No.	Item	Score				
		1	2	3	4	5
1	To encourage managers and R&D personnel of overseas R&D branches to be in regular contact					
2	To encourage managers and management personnel to exchange and communicate regularly with R&D headquarters of the parent company					
3	To hold formal and informal seminars regularly at overseas R&D branches					
4	To create a culture of open information exchange at overseas R&D branches					
5	To organize an international team and encourage researchers from different countries to establish personal relationships among themselves and to influence each other					
6	To rotate jobs at different overseas branches for researchers and management personnel					
7	To organize internal language training at overseas branches, where participants introduce to each other the cultural as well as social background, habits, and so on of different countries					

(24) What are the levels of impact exerted by the following factors on the effective communication of knowledge within an R&D team of your company's overseas branch?

No.	Factor	Score				
		1	2	3	4	5
1	Transparency level of the R&D network					
2	Codification level of tacit knowledge					
3	Trust among members of R&D teams					
4	Costs from increased distance of communication					
5	Confidentiality of knowledge					

6. Policy factors

(25) With regard to innovation internationalization, how important are the following types of national S&T policies to China currently and in the next five years?

No.	Type	Score									
		Current					Next five years				
		1	2	3	4	5	1	2	3	4	5
1	A policy of adapting, imitating, and following										
2	A policy of vigorously acquiring trade secrets from competitors and protecting the confidentiality of that of its own										
3	A policy of international cooperation										
4	To further develop China's specialty areas in S&T and to build up China's specialized and distinctive technological advantages										

(26) How much impact do the following policies of the host country of your company's R&D branches have on your company's R&D and innovative activities?

No.	Policy	Score				
		1	2	3	4	5
1	Intellectual property rights policy					
2	Foreign trade policy					
3	Taxation, financial, and fiscal policies					
4	S&T and innovation policy					

(27) After the accession to the WTO, China will implement intellectual property protection according to the WTO agreement, to what degree do you think this will impact your company?

No.	Item	Score				
		1	2	3	4	5
1	Protection of intellectual property rights will further increase the monopoly of developed countries in knowledge supply					
2	Chinese enterprises will have to obtain foreign technology at a higher cost					
3	Developed countries will slow down their technology transfer to China					
4	To an extent, prevent repetitive development of the functionally same software products and prevent the implementation of reverse engineering					
5	Prevent much of the imitative innovation, and as a result, China has to strengthen R&D to obtain its own intellectual property rights and Chinese enterprises are under tremendous funding pressure of R&D activities					

(28) In order to support enterprise technological innovation and to support enterprise R&D internationalization activities, the country adopts the following policies (to avoid the relevant provisions of the WTO Agreements). How important are such policies to your company?

No.	Policy	Score				
		1	2	3	4	5
1	In technology development, the government increases subsidies for the purchase of patents and other soft technologies and provides a favorable environment for implementing a technology innovation speed-up strategy that is based on imitative innovation					
2	About organizational support and close collaboration among industries, colleges, and research institutes, the government increases subsidies and grants for universities and public research institutes that work in cooperation with business enterprises					
3	The government provides or stimulates market-based financing channels for commercial technological innovation, such as the cultivation of venture capital market, etc.					
4	The government supports R&D indirectly, fosters intermediary R&D units, and creates effective market demand through invitation to tender and government procurement contracts					

7. Process

(29) What are the reasons for failures of technological innovation projects at overseas R&D branches?

No.	Reason	Score				
		1	2	3	4	5
1	Fund shortage					
2	Degree of technological difficulties					
3	The level of technological content and level of commercialization of the company's products are lower than those of its competitors					
4	Difficult to control the complex relationships with its overseas partners					
5	Cultural differences, language differences, and habit differences between Chinese and foreign researchers lead to difficulties in exchange and communication and prevent the formation of effective cooperation and eventually make it difficult for research projects to be smoothly implemented					
6	Domestic parent company and overseas R&D branches cannot carry out effective exchange and communication, and as a result, overseas R&D branches cannot fully implement the technology strategy and technology goals of the parent company					

(30) How well does the actual process match the planned process of innovation projects at your company's overseas R&D branches?

No.	Item	Score				
		1	2	3	4	5
1	Innovation projects are completed within the budget					
2	Innovation projects are completed within the planned time frame					

(31) In an innovation project which necessarily involves multiple overseas R&D units working jointly in development, how important are the following factors?

No.	Factor	Score				
		1	2	3	4	5
1	Division of a large project into subprojects and its implementation into key phases					
2	Setting up a team specialized in coordinating the various R&D institutions					
3	Regular information exchange among R&D institutions					
4	Progress synchronization and uniformity of various R&D institutions in accomplishing their assigned tasks for each phase — research, development, integration, testing, etc.					
5	Frequent job rotation among different functional departments					
6	Cooperation with external laboratories, outsourcing of non-first-class R&D					
7	Whether or not researchers will be moved along when carrying out knowledge transfer among R&D institutions					
8	Involvement of non-R&D departments during project planning phase					
9	Involvement of non-R&D departments during the early stage of project implementation					
10	Involvement of non-R&D departments during the middle or late stage of project implementation					
11	Involvement of non-R&D departments during the market introduction phase					
12	Innovation projects for stimulating consumer demand (that is, for opening up a new market)					
13	Innovation projects to meet consumer demand that has already been activated					
14	Innovation projects for a mature market					

(32) After a joint development project involving several overseas R&D units has succeeded, in the market introduction phase, what model will your company choose and to what degree do you agree with the statements numbered 7–9?

No.	Model/Statement	Score				
		1	2	3	4	5
1	Wholly owned subsidiary					
2	Joint venture					
3	OEM (outsourcing)					

(Continued)

No.	Model/Statement	Score				
		1	2	3	4	5
4	Product export					
5	M&A (controlling a company in an important market by means of M&A and through this company's appliance of the fruits of innovation projects and carrying out production, sales, and so on)					
6	Transfer of the rights to use a patent					
7	Product characteristics are very important when deciding on a market introduction model					
8	Market characteristics are very important when deciding on a market introduction model					
9	Enterprise capabilities are very important when deciding on a market introduction model					

(33) How do you achieve control over joint development involving multiple overseas R&D units?

No.	Method	Score				
		1	2	3	4	5
1	Annual R&D budget					
2	Various R&D units signing written assignment agreement					
3	Controlling in accordance with the reliability requirements or control requirements of a common process (e.g. a production process)					
4	Developing plans as detailed as possible					
5	Standardizing research at various institutions as much as possible					
6	Controlling the entire R&D project through the overseas R&D unit which is in possession of key technologies and thus at the center of the R&D network					

Index

A

acquisition, development, and transfer of technological information sources 203

C

capital investment and allocation 178

centralization 47

centrifugal force for internationalization 26

corporate technology unit 31

D

decentralization 47

determinants of internationalization of technological innovation 25

development process of internationalization of technological innovation 41

E

environmental factor for internationalization of technological innovation 268

external network 6

G

global-for-global 44

global technology unit 30

globalization 3

globally linked 44

H

home-based augmenting 32

home-based exploitation 32

home country's environmental factors 254

host country's environmental factors 264

human resource management 399

I

innovation source 50
integrated R&D network 45
integration of internationalized innovation process 134
internationalization of technological innovation 5
internationalized R&D activity 87

K

knowledge codification 53
knowledge-intensive industry 272

L

local-for-global 111
local-for-local 111
local technology 33
location choice 36
long wave theory 68

M

motivation for internationalization of technological innovation 13

O

organizational form of internationalization of technological innovation 41
overseas technology monitoring point 97

P

paradigm of technological innovation 5
pre-commercial 85

R

resource allocation and integration 175

R&D investment 5
R&D laboratory 41
R&D process 45

S
selection and acquisition of human resources 191
spillover of local technological innovation 267
strategic approach of internationalization of technological innovation 92
strategic framework of internationalization of technological innovation 16
strategic management 72
SPRE model 13

T
technological innovation 3
technological innovation strategy 14
technology scan 43
technology tracking 186
theory of innovation 67
types of internationalization of technological innovation 30